Fatigue of Materials and Structures

Fatigue of Materials and Structures

Application to Damage and Design

Edited by
Claude Bathias
André Pineau

First published 2011 in Great Britain and the United States by ISTE Ltd and John Wiley & Sons, Inc.
Adapted and updated from *Fatigue des matériaux et des structures 3* published 2009 in France by Hermes Science/Lavoisier © LAVOISIER 2009

ISTE Ltd
27-37 St George's Road
London SW19 4EU
UK

www.iste.co.uk

John Wiley & Sons, Inc.
111 River Street
Hoboken, NJ 07030
USA

www.wiley.com

© ISTE Ltd 2011

Library of Congress Cataloging-in-Publication Data

Fatigue des matériaux et des structures. English
  Fatigue of materials and structures : application to damage and design / edited by Claude Bathias, Andre Pineau.
    p. cm.
  Includes bibliographical references and index.
  ISBN 978-1-84821-267-1
  1. Materials--Fatigue. I. Bathias, Claude. II. Pineau, A. (André) III. Title.
  TA418.38.F3713 2010
  620.1'126--dc22

                                                                    2010040728

British Library Cataloguing-in-Publication Data
A CIP record for this book is available from the British Library
ISBN 978-1-84821-267-1

Printed and bound in Great Britain by CPI Antony Rowe, Chippenham and Eastbourne.

MIX
Paper from
responsible sources
FSC
www.fsc.org   FSC® C013604

Table of Contents

Foreword

This book on fatigue, combined with two other recent publications edited by Claude Bathias and André Pineau[1], are the latest in a tradition that traces its origins back to a summer school held at Sherbrooke University in Quebec in the summer of 1978 which was organized by Professors Claude Bathias (then at the University of Technology of Compiegne, France) and Jean Pierre Bailon of Ecole Polytechnique, Montreal, Quebec. This meeting was held under the auspices of a program of cultural and scientific exchanges between France and Quebec. As one of the participants in this meeting, I was struck by the fact that virtually all of the presentations provided a tutorial background and an in-depth review of the fundamental and practical aspects of the field as well as a discussion of recent developments. The success of this summer school led to the decision that it would be of value to make these lectures available in the form of a book which was published in 1980. This broad treatment made the book appealing to a wide audience. Indeed, within a few years, dog-eared copies of "Sherbrooke" could be found on the desks of practicing engineers, students and researchers in France and in French-speaking countries. The original book was followed by an equally successful updated version that was published in 1997 which preserved the broad appeal of the first book. This book represents a part of the continuation of the approach taken in the first two editions while providing an even more in-depth treatment of this crucial but complex subject.

It is also important to draw attention to the highly respected "French School" of fatigue which has been at the forefront in integrating the solid mechanics and materials science aspects of fatigue. This integration led to the development of a deeper fundamental understanding thereby facilitating application of this knowledge

1. C. BATHIAS, A. PINEAU (eds.), *Fatigue of Materials and Structures: Fundamentals*, ISTE, London and John Wiley & Sons, New York, 2010.
C. BATHIAS, A. PINEAU (eds.), *Fatigue of Materials and Structures: Application to Design*, ISTE, London and John Wiley & Sons, New York, 2011.

to real engineering problems from microelectronics to nuclear reactors. Most of the authors who have contributed to the current edition have worked together over the years on numerous high-profile, critical problems in the nuclear, aerospace, and power generating industries. The informal teaming over the years perfectly reflects the mechanics/materials approach and, in terms of this book, provides a remarkable degree of continuity and coherence to the overall treatment.

The approach and ambiance of the "French School" is very much in evidence in a series of bi-annual international colloquia. These colloquia are organized by a very active "fatigue commission" within the French Society of Metals and Materials (SF2M) and are held in Paris in the spring. Indeed, these meetings have contributed to an environment which fostered the publication of this series.

The first two editions (in French), while extremely well-received and influential in the French-speaking world, were never translated into English. The third edition was recently published (again in French) and has been very well received in France. Many English-speaking engineers and researchers with connections to France strongly encouraged the publication of this third edition in English. The current three books on fatigue were translated from the original four volumes in French[2] in response to that strong encouragement and wide acceptance in France.

In his preface to the second edition, Prof. Francois essentially posed the question (liberally translated), "Why publish a second volume if the first does the job?" A very good question indeed! My answer would be that technological advances place increasingly severe performance demands on fatigue-limited structures. Consider, as an example, the economic, safety and environmental requirements in the aerospace industry. Improved economic performance derives from increased payloads, greater range and reduced maintenance costs. Improved safety, demanded by the public, requires improved durability and reliability. Reduced environmental impact requires efficient use of materials and reduced emission of pollutants. These requirements translate into higher operating temperatures (to increase efficiency), increased stresses (to allow for lighter structures and greater range), improved materials (to allow for higher loads and temperatures) and improved life prediction methodologies (to set safe inspection intervals). A common thread running through these demands is the necessity to develop a better understanding of fundamental fatigue damage mechanisms and more accurate life prediction methodologies (including, for example, application of advanced statistical concepts). The task of meeting these requirements will never be completed; advances in technology will require continuous improvements in materials and more accurate life prediction schemes. This notion is well illustrated in the rapidly developing field of gigacycle

---

2. C. BATHIAS, A. PINEAU (eds.), *Fatigue des matériaux et des structures*, Volumes 1, 2, 3 and 4, Hermes, Paris, 2009.

fatigue. The necessity to design against fatigue failure in the regime of 10^9 + cycles in many applications required in-depth research which in turn has called into question the old, comfortable notion of a fatigue limit at 10^7 cycles. New developments and approaches are an important component of this edition and are woven through all the chapters of the three books.

It is not the purpose of this preface to review all of the chapters in detail. However, some comments about the organization and over-all approach are in order. The first chapter in the first book[3] provides a broad background and historical context and sets the stage for the chapters in the subsequent books. In broad outline, the experimental, physical, analytical and engineering fundamentals of fatigue are developed in this first book. However, the development is done in the context of materials used in engineering applications and numerous practical examples are provided which illustrate the emergence of new fields (e.g. gigacycle fatigue) and evolving methodologies (e.g. sophisticated statistical approaches). In the second[4] and third[5] books, the tools that are developed in the first book are applied to newer classes of materials such as composites and polymers and to fatigue in practical, challenging engineering applications such as high temperature fatigue, cumulative damage and contact fatigue.

These three books cover the most important fundamental and practical aspects of fatigue in a clear and logical manner and provide a sound basis that should make them as attractive to English-speaking students, practicing engineers, and researchers as they have proved to be to our French colleagues.

Stephen D. ANTOLOVICH
Professor of Materials and Mechanical Engineering
Washington State University
and
Professor Emeritus
Georgia Institute of Technology

December 2010

---

3. C. BATHIAS, A. PINEAU (eds.), *Fatigue of Materials and Structures: Fundamentals*, ISTE, London and John Wiley & Sons, New York, 2010.
4. C. BATHIAS, A. PINEAU (eds.), *Fatigue of Materials and Structures: Application to Design*, ISTE, London and John Wiley & Sons, New York, 2011.
5. This book.

Chapter 1

High Temperature Fatigue

1.1. Introduction and overview

1.1.1. *Introductory remarks*

It is a basic consequence of thermodynamics that the efficiency of heat engines, regardless of their type, increases with increasing temperature. In the power generation industry (nuclear industry, coal-fired and/or oil-fired plants), any increase in working temperature leads to a decrease in fuel consumption, pollution and operating costs. In the jet engine industry, increased operating temperatures lead to improved performances, such as the combinations of heavier payloads, a greater speed and a greater range.

For the power generation industry, lower fuel consumption, reduced pollution and lower costs are important. However, as operating temperatures are increased, additional problems to those encountered at lower temperatures arise. Regardless of its type, all engines have moving parts that experience variable loading during each operating cycle. In general, loading above a certain level causes microscopic rearrangements at the atomic level, which can lead to an important damage. With continued operation (i.e. cyclic loading) damage accumulates and eventually leads to the fracture of the component. This scenario can be viewed as a working definition of fatigue. Many practical and theoretical investigations have been carried out over the past two centuries to experimentally characterize failure by fatigue and to predict the lives of components subjected to fatigue loading. Wöhler [WÖH 1860], going back more than a century, demonstrated that the fatigue life of a component may be

Chapter written by Stephen D. ANTOLOVICH and André PINEAU.

represented in terms of stress, which eventually leads to the well-known and widely used S/N curve methodology, which was discussed in Claude Bathias and André Pineau's *Fatigue of Materials and Structures: Fundamentals*. Even if fundamental and practical difficulties are still encountered using this approach, the stress-based method is applicable to this day to make preliminary life estimates of some components.

1.1.2. *A little history*

1.1.2.1. *Effects of temperature*

At higher temperatures such as those found in the power generation and jet engine industries, not only is there repeated loading, but depending on strain rates and hold times, time-dependent damage processes such as creep and environmental attack become important. Two major advances in understanding high temperature fatigue from an engineering perspective were made in the 1950s and 1960s at NASA Lewis by Manson and coworkers (see e.g. [MAN 53]) and at General Electric by Coffin and coworkers [COF 54]. The first was in terms of conceptualizing the fatigue process in a more physically acceptable, albeit very general, manner. The second major advance was associated with advances in control technology. The rapid advances in "controls" paved the way for the development of closed-loop test machines, much more reliable data, and vastly improved life prediction capabilities. These are discussed in greater detail below.

1.1.2.2. *Problems with the S/N approach: strain as a driver of damage*

Attempts to predict the lives of high-temperature components in jet engines and nuclear power generation facilities using the S/N approach lacked sufficient precision and required the use of significant safety factors. Two factors limited the utility of the traditional approaches:

– dynamic changes in the material (intrinsic and extrinsic) are not captured in the traditional S/N approach (this aspect is a major focus of this chapter and is discussed in detail in subsequent sections).

– stress is not a physically meaningful damage function in itself.

The second limitation is easily understood in terms of basic physics. It is not possible to "measure" a stress (or equivalently a force) without some reference to displacement and a relationship defining stress (or force) in terms of strain (or displacement). Strains and displacements are real, while stresses and forces represent mathematical conveniences. Real damage in a material will always depend upon some kind of displacement or physical change within the material.

In a structural crystalline material, it is intuitively appealing to associate damage with strain. The more strain, the more dislocation movement. The more dislocation movement, the greater the possibility of dislocation intersections and pile-ups. These obviously produce various forms of damage such as vacancies, interstitials, cells, tangles and increased dislocation densities (see e.g. [HUL 01] for a detailed discussion of dislocation interactions and debris). In fatigue, we are thus led to think in terms of the reversed plastic strain, $\Delta\varepsilon_p$. In addition, elastic strains, denoted by $\Delta\varepsilon_e$, can be damaging in the sense that they may produce microplasticity or elastic displacements that may be sufficient to cause cracking of brittle inclusions. Both Coffin [COF 54] and Manson [MAN 53], working independently, adopted the working hypothesis that plastic strain range was the real driver of fatigue damage and hence determined the fatigue life. Using the then recently developed closed-loop test equipment they were able to carry out experiments in which the strain range was controlled. For relatively high strains, they found that when the plastic strain range was plotted against the cyclic life, a straight line on a log-log plot was produced, independent of material. The result was the well-known Coffin-Manson Law:

$$N_f^{\beta} \cdot \Delta\varepsilon_p = C_D \qquad\qquad [1.1]$$

These terms are defined in [BAT 10] and will not be discussed here, other than to emphasize that the exponents and constants are dependent upon the material and testing conditions. Most of the currently used high-temperature life prediction methodologies use equation [1.1] as a starting point and modify it in some way to account for time-dependent processes, such as creep and environment effects.

1.1.2.3. *Total strain approach to life prediction – advantages and critique*

Manson [MAN 65] and his coworkers [HAL 78] at NASA recognized that putting life in terms of total strain was more convenient from an engineering perspective. This was done by noting that the total strain range, $\Delta\varepsilon_t$, was the sum of the elastic and plastic ranges and that a correlation similar to equation [1.1] could be made. Adding up the elastic and plastic strain ranges gives:

$$\Delta\varepsilon_t = C_D \cdot N_f^{-\beta} + C_E \cdot N_f^{-\alpha} \qquad\qquad [1.2]$$

The second set of quantities on the right-hand side corresponds to the elastic strain component. For many materials, the values of β and α are approximately 0.6 and 0.12 [PIN 10] respectively. However, since these are only approximate values, their use is limited to scoping calculations. Due to the form of equation [1.2] and the values of β and α, the first component dominates for high strains and the second component dominates for low strains. Thus, life prediction may be viewed in terms of low cycle fatigue (high strain ranges) or high cycle fatigue (low strain ranges). Only when the elastic and plastic strains are essentially equal must we consider both terms.

While the strain-life equation [1.2] is frequently used, there are two problems with this equation and with equation [1.1]. The first one is that life is traditionally represented in terms of cycles to failure, N_f . Clearly this introduces geometric dependence into the problem and implicitly incorporates a process that was not addressed – the process of crack propagation. Many researchers have attempted to address this situation by using cycles to initiation, defined as a given load drop in a strain-controlled test. The relationship between load drop and crack size is easily developed. It is given by:

$$A_c = 100(\%LD)A_0$$
[1.3]

where A_c is area of the crack, A_0 is the initial cross-sectional area of the sample, and $\%LD$ is the percentage load drop.

With a specimen 6.35 mm in diameter, a 1% load drop corresponds to a crack with a characteristic linear dimension of about 560 µm[1]. For most polycrystalline materials, this corresponds to at least 10 grains and it is difficult to consider a crack of such a size as relating entirely to initiation. The size of the crack detected via this method may be decreased by reducing both the load drop and the specimen's diameter. Clearly distinguishing the initiation phase (e.g. limiting initiation to a physically meaningful dimension) remains a challenge in high temperature fatigue.

The second problem is that by representing the fatigue life (or more correctly the cycles to initiation) by equation [1.2], the dependent and independent variables are confounded in such a way that equation [1.2] cannot be restructured to make life a function of strain; a more rational dependence. Equation [1.2] can easily be solved using numerical iteration techniques. It does not present any difficulty from an engineering perspective. However, the form of the equation does present scientific difficulties. A fundamental model of fatigue would have to have the following form:

$$N_i = f(D_{\Delta\varepsilon_p}(N))$$
[1.4]

where N_i is the number of cycles to initiation, D is the damage state which depends on the current cycle number for given plastic strain range, $\Delta\varepsilon_p$, and N is the current cycle number.

Equation [1.2] and its variants should be regarded as an engineering convenience and not as a scientific basis for a deep understanding of fatigue.

---

1. This assumes that the material is stable and that softening does not occur. It is considerably more difficult to distinguish load drops from cracking from those due to softening.

1.1.3. *High temperature testing closed-loop control and extensometry*

The closed-loop approach to mechanical testing, in its simplest form, means that what is desired is actually delivered. This topic is discussed in more detail elsewhere [ANT 00] but a brief description is important here, not only for the historical context, but for its centrality in high temperature low cycle fatigue (LCF). In broad outline, a program for stress or strain as a function of time is programmed for a specimen. The closed-loop control system compares the command value to the measured value and makes a rapid adjustment to bring the system into dynamic equilibrium. In this way, we may obtain test results with a high degree of accuracy and precision for precisely controlled conditions. In high temperature studies, it is customary to use plastic strain control since it directly affects damage. In practice, the upper and lower plastic strain limits are set along with the loading rates and hold times (i.e. the loading profile) and the test is conducted in this way. The appeal of this approach is that it corresponds to the real physical quantity, which is closely related to damage.

Implicit in this approach is the ability to measure and control temperature, which can also be done using a feedback system with inductive or radiative heating. Some care must be exercised here to avoid convective effects and to make sure that thermocouples, if used with inductive heating, do not have a significant loop cross section normal to the flux lines lest extraneous EMF (Electromagnetic force) be generated. High-temperature extensometry must be employed, adding a further complication. If mechanical contact is made between the specimen and the extensometer, such contact must be firm enough to prevent the extensometer from slipping (and causing undesired loading) but not so firm as to cause crack initiation and premature fatigue failures at the point of extensometer contact. Extremely simple, cost-effective and accurate high-temperature extensometers have been developed that completely avoid these problems [MIL 88]. Another method involves the use of laser extensometry. Due to changes in the surface at high temperatures and slower response rates, however, lasers are best used for measuring rather than controlling. In addition, laser extensometers are at least 30 times more expensive than the mechanical extensometers discussed above and are not easily replaced.

1.1.4. *Damage mechanisms and interactions in high-temperature fatigue*

In addition to the damaging effects of reversed plasticity, creep and environmental effects (usually in the form of oxidation) must be considered at high temperatures. Furthermore, interactions among these mechanisms must be measured as well as the stability of the underlying microstructure and dislocation substructure. As a simple example, slip bands may impinge upon oxygen-poisoned boundaries causing early cracking. In addition, the ingress of oxygen atoms may affect the way in which fundamental fatigue deformation processes occur. Complicating matters further is the fact that in many real-world applications, the mechanical strains and temperature vary independently (so-called thermomechanical fatigue, or TMF, discussed in more detail in Chapter 7 of [BAT 11]) leading to damage mode interactions that probably never occur in isothermal fatigue. Any physically meaningful life prediction equation must be such that its form and physical parameters correspond to physical reality. This aspect of life prediction is discussed in detail in this chapter.

1.1.5. *Organization of this chapter*

This chapter starts out with an overview of several existing models for high-temperature fatigue to provide background and context. High-temperature fatigue is a vast and complex problem in which many materials are considered. Such materials are so vast in number that an intelligent limitation must be made while still elucidating important principles. In order to do this, we have decided to use the following three important and representative classes of material to achieve this goal:

– austenitic stainless steels (power generation industry);

– Cr-Mo steels (power generation industry);

– Ni-base superalloys (gas turbines in the jet engine industry).

In addition to their industrial importance, these materials were selected for in-depth discussion because their different microstructures, deformation characteristics, and damage mechanisms fully illustrate the challenges of high-temperature fatigue. With some caution, the principles developed in this chapter may be applied to other material systems. In the section on each of these materials, the basic microstructure, fundamental deformation mechanisms, basic fatigue damage mechanisms and their interactions are discussed. In some cases, TMF is also included. Models based on dominant mechanisms are presented in several cases. Some of the more widely used engineering models are discussed with respect to the physics-based models (which are the main focus of this chapter) in terms of functional form and sensitivity of parameters to temperature, strain rate and environment.

An important theme running throughout this chapter, for both LCF and fatigue crack propagation (FCP), is the dynamic nature of materials with respect to the stability of the microstructure and dislocation substructure as a function of temperature, level of deformation and strain rate.

1.1.6. *Goals*

The goals of this chapter are:

– to review mechanisms of cyclic deformation, damage accumulation and crack propagation in austenitic stainless steels, Cr-Mo steels, and superalloys (both Ni-base and Fe-Ni base);

– to relate these mechanisms to engineering applications;

– to review life prediction methodologies appropriate to these applications and mechanisms;

– to point out current and likely future trends in the development of more fatigue-resistant materials and improved life prediction methodologies.

1.2. 9 to 12% Cr steels

1.2.1. *Introduction*

Martensitic steels of the 9–12% Cr variety are widely used in the thermal power generation industry where operational temperatures are in the neighborhood of 550°C. The desire for increased operating temperatures led to the development of compositional modifications in these materials. The first of these modifications was the addition of 1% molybdenum followed by tungsten, vanadium and nitrogen. Most of the structures in this industry are subject to creep loading. Table 1.1 summarizes the various compositions that have been developed and gives the 600°C creep rupture stress. Considerable hope is vested in the most advanced compositional modifications for the fabrication of components to be used in fourth-generation electro-nuclear fusion reactors scheduled to come on line around 2030–40. This timeframe requires a demonstration reactor to be available around 2020. This means that most technical solutions must be well in hand around 2012. In the same vein, we can also consider the possibility of using these steels for fusion reactors whose industrial debut is further in the future.

Compared to austenitic stainless steels, these steels in addition to other advantages have superior thermal conductivity, a lower coefficient of thermal expansion and less sensitivity to radiation-induced swelling, especially the ODS (Oxide Dispersion Strengthened) versions that are produced by mechanical grinding using ball milling. To design these reactors, it is imperative to have a detailed understanding of the fatigue and creep-fatigue properties of these steels since cycling is related to maintenance-related shutdowns and startups, and to a lesser extent on the demands of the power grid. The major problem associated with the design rules for components used in the nuclear power generation industry lies in the necessity of extrapolating the data upon which these rules are based. The totality of these technical requirements explains the renewed interest in studying these materials in the past few years.

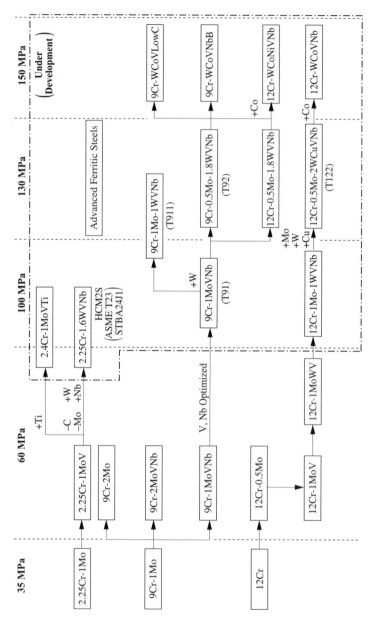

Table 1.1. *Chromium-bearing ferritic and martensitic steels showing the evolution in chemistry and improvement in 100,000-hour creep strength at 600°C*

In this chapter, a classical approach to considering fatigue and creep fatigue properties is followed. After a discussion of the microstructure, classical fatigue behavior is considered. It is shown that the particularities of the microstructure lie at the heart of the important phenomenon of cycling softening that is often observed. Next, the subject of fatigue damage is considered. The discussion of steels is centered on those with compositions of 9% Cr and 1% Mo (P91). However, changes in behavior associated with compositional modifications and microstructural changes are discussed, as far as is possible. A number of excellent recent review papers are available on these steels. For example, the review of Klueh and Nelson [KLU 07] can be consulted on this topic. This section relies heavily upon the thesis by Fournier [FOU 07], which was the basis of several recent publications: [FOU 06a, FOU 06b, FOU 08a, FOU 08b, FOU 08c, FOU 09a, FOU 09b]. Fournier studied a composition characteristic of these steels (C = 0.088, N = 0.043, Cr = 8.78, Mo = 0.92, Mn = 0.35, Si = 0.33, Nb = 0.78, V = 0.191 in percentage weight). This steel was austenitized at 1,050°C for 30 minutes, quenched and then aged at 780°C for one hour. The results obtained on this steel serve as the basis for the following discussion. The results for other steels in this family, notably those containing 12% Cr, are discussed.

1.2.2. *Microstructures of 9–12% Cr steels*

The microstructure of these steels must be considered at several size scales, see Figure 1.1.

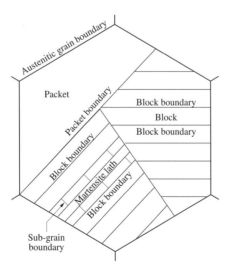

Figure 1.1. *Various significant size scales for quenched and tempered 9–12% Cr steels [FOU 07]*

At the optical microscopy level, entangled packets resulting from the martensitic phase transformation can be seen in prior austenite grains (d~10–60 μm). These packets are subdivided into blocks of laths, which have the same $\{111\}_\gamma$ habit plane.

The laths have a variable thickness but are generally in the order of several microns. During quenching, subgrains are formed in the interior of these laths (see Figure 1.2). These subgrains also have a variable size but are generally in the order of about 500 nm. The microstructure and crystallography of the martensite lath obtained from quenching these steels is very well described in the article by Kitahara *et al.* [KIT 06].

All of the steels have carbides after ageing whose dimensions vary significantly with composition and the thermal treatment. Chromium carbides of the $M_{23}C_6$ type are the first to form and are quite often preferentially precipitated upon boundaries, including former γ grain boundaries and others (see e.g. [EFF 89]).

Figure 1.2. *P91 steel. Transmission electron microscopy observations:
a) initial condition (quenched and tempered); b) cyclically deformed in pure fatigue
($\Delta\varepsilon_{fat}$ = 0.70%), T = 550°C; c) tested in hold-time fatigue ($\Delta\varepsilon_{fat}$ =0.70%), T=550°C.*

Note the progressive coarsening of the structure from a) to c) [FOU 88]

Moreover, these steels contain very fine carbides (and carbo-nitrides) of the MX type that are distributed uniformly throughout the microstructure. More details are provided elsewhere (see, e.g. [GAF 05] and [BRA 91]).

1.2.3. *Mechanical behavior*

In the following, the mechanical behavior of 9–12% Cr steel during fatigue and creep-fatigue are discussed.

1.2.3.1. *Behavior in pure fatigue*

Typical fatigue behavior for a 9–12% Cr steel is shown in Figure 1.3 [ARM 04]. It is seen that the maximum stress decreases continually at both low ($\Delta\varepsilon_t$ = 0.6%) and high ($\Delta\varepsilon_t$ = 1.0%) strains. This softening becomes more pronounced as the temperature increases. Thus, at T=550°C and at half of the fatigue life, the stress is practically cut in half. A detailed study by Fournier [FOU 07] on a P91 steel showed that this cyclic softening seen at 550°C (Figure 1.4), affects the kinematic component of hardening whereas the isotropic component is hardly affected over the strain range that was studied (0.4%< $\Delta\varepsilon_t$ <0.6%) [FOU 06]. This observation suggests that microstructural modifications occur during cycling that affect the long-range internal stresses associated with the arrangement of dislocations.

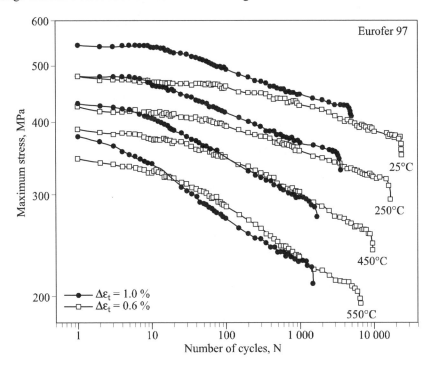

Figure 1.3. *Eurofer 97 steel. LCF tests at various temperatures (25°C, 250°C, 450°C and 550°C) at two strain levels – $\Delta\varepsilon_t$ =0.60% and 1%. The variation of stress with the number of cycles is shown [ARM 04]*

Transmission electron microscopy (TEM) examinations revealed a significant modification of the dislocation substructure, especially when a creep component is imposed on a pure fatigue test, Figure 1.2. The density of dislocations decreases and dislocations re-arrange themselves into larger cells. The relationship between the cell size and saturation stress (or the stress at half-life) is shown in Figure 1.5 [KLM 88]. It thus appears as if the microstructure of 9–12% Cr steels is highly unstable under cyclic loading. Recent studies on steels produced by ball milling systematically seem to show that the presence of very fine particles in the matrix strongly reduce or even suppress this softening phenomenon [UKA 07].

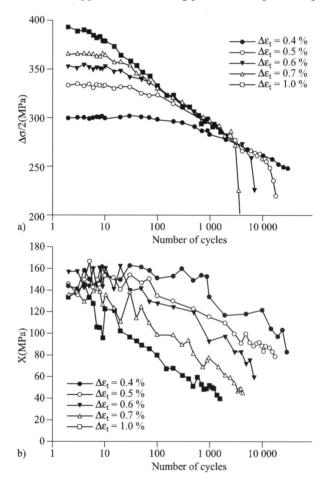

Figure 1.4. *P91 steel: a) variation in the stress amplitude with the number of cycles in continuous fatigue; b) variation of the kinematic component, X, of hardening with the number of cycles. The total applied strain amplitude at 550°C is indicated on each figure [FOU 06]*

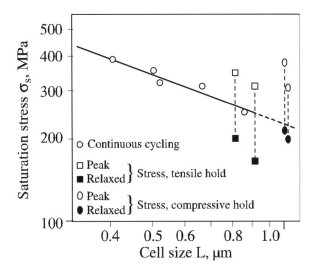

Figure 1.5. *Correlation between cyclic softening and microstructural coarsening for a P91 steel fatigued at 593°C [KIM 88]*

1.2.3.2. *Creep fatigue behavior*

The principle of creep-fatigue testing with a hold time is shown in Figure 1.6. In relaxation fatigue tests, Figure 1.6a, a hold is applied at the maximum strain (tension) or the minimum strain (compression). During this hold period, stress relaxation occurs by a mechanism of viscous deformation process in which elastic strain is converted into plastic strain.

In an actual creep-fatigue test a hold time is imposed at the maximum (or minimum) stress, which is held constant, until the desired level of strain is attained, see Figure 1.6b. During this hold time, a creep strain, ε_{creep}, occurs either in tension or in compression. These creep-fatigue tests are more revealing of the behavior, in as much as an increasing amount of creep strain is applied on a cycle-by-cycle basis. These tests lead to the development of a mean stress in tension for a compressive hold and a compressive mean stress for a tensile hold, as seen in Figure 1.7. A very strong softening effect is seen in Figure 1.7.

It has been possible to show that for a test in which $\Delta\varepsilon_{fat}$ is controlled, this progressive softening as measured by the change in the stress amplitude is uniquely a function of the viscous flow. Results obtained for $\Delta\varepsilon_{fat} = 1\%$ are seen in Figure 1.8.

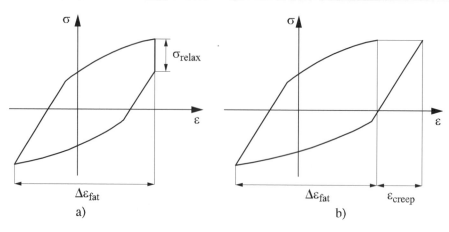

Figure 1.6. *Schematic form of stress-strain hysteresis loops: a) fatigue-stress relaxation; and b) creep-fatigue [FOU 06]*

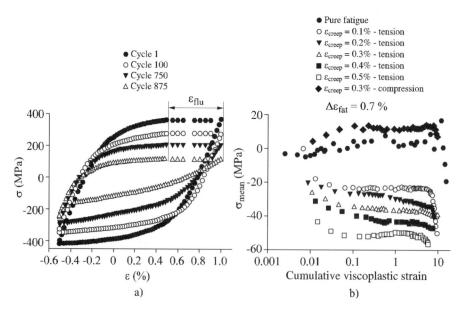

Figure 1.7. *P91 steel: a) shapes of hysteresis loops for a creep-fatigue test with $\Delta\varepsilon_{fat} = 1\%$ and $\varepsilon_{creep} = 0.5\%$; b) variation of the mean stress during continuous cycling and creep-fatigue with hold times in tension and one test with a compression hold [FOU 06]*

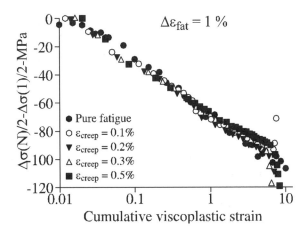

Figure 1.8. *P91 steel. Stress amplitude at half-life as a function of total accumulated creep strain. Testing done at 550°C with a fatigue strain amplitude of 1% [FOU 06]*

TEM analysis showed that the phenomenon of coarsening of the structure at the dislocation level is particularly pronounced (see e.g. Figure 1.2c). The size distributions of subgrains in the initial state, in the pure fatigue state and in the creep-fatigue state are shown in Figure 1.9a, while the distribution of average misorientations between blocks of laths for these states is shown in Figure 1.9b.

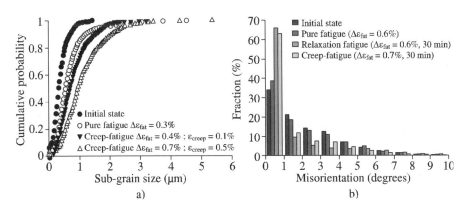

Figure 1.9. *P91 steel: a) variation of the distribution of subgrain sizes as measured by transmission electron microscopy; b) distribution of average misorientations between blocks of laths. The test conditions at 550°C are indicated [FOU 06]*

Note that the coarsening of the microstructure is more pronounced in creep-fatigue than in pure fatigue. The misorientation histograms show that after creep-fatigue or stress relaxation fatigue, the fraction of slightly misoriented (θ <1 degree) is two times as large as for the initial state. This illustrates the fact that the blocks are more homogeneous, which is associated with some of them being eliminated.

1.2.4. *Damage*

1.2.4.1. *Continuous cycling fatigue*

Results showing the fatigue life in terms of total strain are presented in Figure 1.10 for several temperatures. In this representation, increasing the temperature causes, on average, a reduction in the cyclic life. This reduction is observed for P91 steel as well as for a 12% Cr steel.

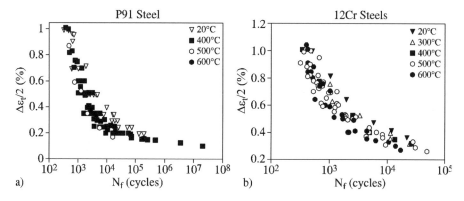

Figure 1.10. *Influence of temperature on the fatigue life of two different compositions of martensitic steel in terms of the total strain amplitude*

Since, as previously shown, these materials undergo marked cyclic softening, we might legitimately ask whether the same marked effects of temperature are seen when the data are represented in terms of a Coffin-Manson plot.

This is done in Figure 1.11, where it can be seen that on average the effect of temperature on fatigue life is reduced. This once again illustrates the importance of taking the cyclic softening into account.

The observed softening is associated with an increase in the effective ductility and a reduction in stress for plastic strain control. The reduced stress lowers the driving force for both crack initiation and extension.

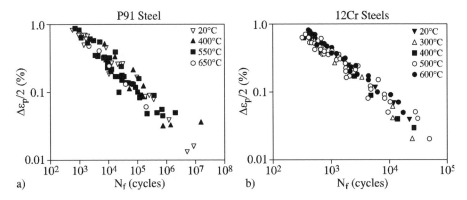

Figure 1.11. *Influence of temperature on the fatigue life of two martensitic steels as a function of the applied plastic strain amplitude [FOU 08]*

1.2.4.2. *Effects of hold time*

The effect of tensile hold time during creep-fatigue testing of P91 steel is shown in Figure 1.12 [FOU 08]. We can see that the imposition of a creep component to a fatigue test leads to a significant reduction in lifespan. A detailed study confirmed that P91 steel, like all others in this class, is more sensitive to a compression hold than to a tensile hold. This negative effect of a compression hold is increasingly marked at low fatigue strain amplitudes, see Figure 1.13. This effect has previously been seen in lower Cr-bearing steels of the 2 ¼ Cr-1Mo type (see e.g. [CHA 81]).

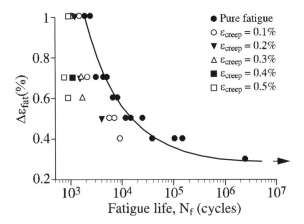

Figure 1.12. *P91 steel. Influence of a tensile hold (creep) on the creep-fatigue life at 500°C compared to results for continuous cycling. The creep deformation strains are indicated [FOU 08]*

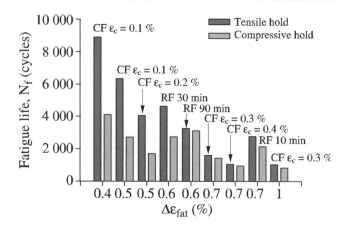

Figure 1.13. *P91steel. Life as a function of fatigue strain range for various levels of imposed creep strain in tension and compression. The values of creep strain are indicated in the figure [FOU 07]*

Several hypotheses can explain these hold time effects, especially in compression, on the creep-fatigue life of these steels. The first is of a purely mechanical nature. A compressive hold may be all the more damaging as it leads to a tensile mean stress. However, this is probably a second-order effect since in a vacuum there is no significant effect between tensile and compressive holds [FOU 08]. We must thus look for an explanation in terms of oxidation.

A detailed study of damage in P91 tested at 550°C was carried out by Fournier [FOU 07, FOU 08]. No significant intergranular damage was observed in any case. However, it was clearly shown that hold times (in tension or compression) profoundly affected crack initiation. Specimens tested in air always had an oxide on the surface whose thickness increased with increasing hold time. For thin oxide layers, that is for short hold times and low strains, crack initiation was the same as in non-oxidizing environments and occurred in a classical Stage I fashion. For thicker oxide layers (i.e. longer hold times and higher strain levels) the consequence was brittle cracking in the oxide. These propagated as Stage II cracks. For the definition of Stages I and II see Chapter 4 of [BAT 10].

This difference in the fatigue crack initiation mechanisms leads quite naturally to the notion that two regions of behavior exist that are separated by a transition; see Figure 1.14. In region I, oxidation-related phenomena play a small role. Nevertheless, it has been shown that the fatigue life in vacuum is about three times longer than in air. In region II, initiation takes place by brittle fracture of the oxide layer and is accelerated by internal oxidation in the crack tip region. The transition between these two regions has been defined as shown in Figure 1.14. It remains a challenge to predict the exact form of this transition between the two regions. The position of the transition for a compressive hold time is also represented in this figure. For a given hold time, the transition occurs at a lower strain level, in qualitative agreement with the corresponding lives.

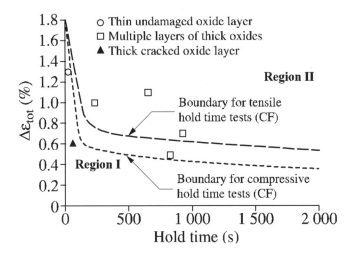

Figure 1.14. *Steel P91. Definition of two regions of behavior corresponding to different mechanisms of crack initiation. The total strain range is defined by* $\Delta\varepsilon_{tot} = \Delta\varepsilon_{fat} + \Delta\varepsilon_{creep}$.

The boundary between the two regions varies with hold time (tension or compression). For the precise definition of regions I and II, see [FOU 08]

The observed difference between tensile and compressive holds can easily be explained in a qualitative manner using the schematic in Figure 1.15. For compressive holds, the oxides that form during the hold are essentially stress free. As soon as deformation starts in the tension-going direction, the oxide is in tension even though the body as a whole is in compression. Thus, the brittle oxide is in tension for the full strain range, favoring the brittle fracture of this layer.

When the oxides form during a tensile hold, they are placed into compression as soon as the compression-going part of the cycle starts and the oxide is placed in a state of compression corresponding to the full strain range. This favors oxide spalling as opposed to crack formation in a compressive hold.

Quantitative definition of the transition between the two regions is difficult since it rests on a calculation that requires properties of the oxide that are not generally available.

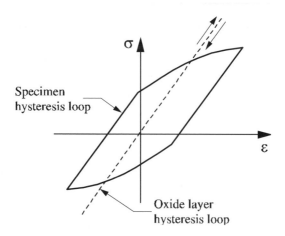

Figure 1.15. *Schematic stress-strain diagram showing the stress-strain path that the oxide undergoes during a fatigue cycle [FOU 08]*

1.2.5. *Damage model and life prediction*

Two regions of behavior have been identified for crack initiation. It is now possible to develop a physically-based model for life prediction. In doing this, the classical notion of dividing fatigue into an initiation stage (N_i) and a propagation stage (N_p) is used:

$$N_f = N_i + N_p \hspace{4cm} [1.5a]$$

Depending on whether the strain range and hold time place the part of interest in region I or II, the initiation phase can be schematically represented as follows:

– In region I, the model proposed by Mura [MUR 94] and Tanaka and Mura [TAN 81] (also see Brückner-Foit and Huang [BRU 06] and Alexandre *et al.* [ALE 04a, ALE 04b]), may be expressed as:

$$N_i = \frac{\alpha}{d}\frac{1}{\Delta\varepsilon_p^2} \hspace{4cm} [1.5b]$$

where d is the "grain" size, assumed in this case to be 20 μm, and $\Delta\varepsilon_p$ is the plastic strain range. The constant of proportionality α has been calculated from an extensive data set taken at $\Delta\varepsilon_{tot} = 0.40\%$ to be 0.0385.

– In region II, fracture of the oxide layer takes place very quickly and the number of cycles to initiation can be taken to be zero.

For the two types of damage, crack propagation is well described by an equation proposed by Tomkins [TOM 68], which is:

$$\frac{da}{dN} = \frac{\pi^2}{8} \frac{\Delta\varepsilon_p \Delta\sigma^2}{\left(2\overline{T}\right)^2} a \left[1 + \frac{\pi^2}{8}\left(\frac{\Delta\sigma}{2\overline{T}}\right)^2\right] \qquad [1.6]$$

where a is the length of the microcrack.

This expression predicts that the crack length varies exponentially with the number of cycles. Experiments carried out on LCF specimens containing small cracks (0.5 mm) validated this expression. The only adjustable parameter is $\overline{T}$ which, for this data set, could be taken as approximately 680 MPa. This value appears to be high, considering that the strength of this material in tension is only 380 MPa.

The number of cycles in crack propagation is obtained by integration of equation [1.6] between two limits:

$$N_p = \int_{a_o}^{a_c} \frac{da}{da/dN} \qquad [1.7]$$

where a_o is the initial crack size and a_c is the crack size at which the specimen may be considered to be fractured. In this case, a_c was taken as the radius of the specimen and a_o was taken as a fixed value of 50 μm. See the work of Fournier [FOU 08] for a detailed discussion of the choice of the initial crack length, which is very important for all models of this type.

Another difficulty in applying models of this type is related to the strong cyclic softening effect, which brings some uncertainty to the values of $\Delta\sigma$ and $\Delta\varepsilon_p$ that are used in equation [1.6] since they change continuously during the test.

Finally, experience has shown that in the case of type II damage, there are generally five cracks in the fracture plane. Thus the value of a_o is divided by five. The results obtained for the ensemble of pure fatigue and creep-fatigue tests are shown in Figure 1.16. The values of $\Delta\sigma$ and $\Delta\varepsilon_p$ were taken at the half-life; Figure 1.16. It can be seen that on the whole this simple model predicts the life in a satisfactory way.

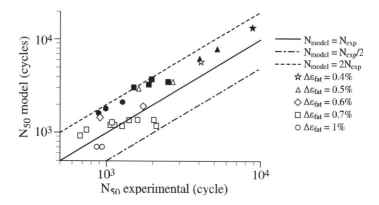

Figure 1.16. *P91 steel. Comparison between actual and predicted life. Values of*
$\Delta\sigma$ *and* $\Delta\varepsilon_p$ *were measured at* N_{50} *. The open symbols correspond to type II damage,*
whereas the filled symbols correspond to type I damage. The strain amplitudes in fatigue
are indicated on the graph [FOU 07]

The agreement that was observed reinforces the notion of two types of damage for this kind of steel in creep-fatigue loading. It is not due to massive intergranular damage, which would profoundly modify the rate of crack propagation. Rather it is due to the microstructural instability that engenders cyclic softening and, beyond that, in the creep-fatigue regime to accelerated crack initiation brought about by fracture of the surface oxide film.

This conclusion leads to the idea that if improvements in creep-fatigue behavior are sought, the most promising path would be to develop alloys with improved microstructural stability. For example, stable structures can be obtained in existing alloys through thermo-mechanical treatments that refine the carbides and disperse them more homogeneously. This may be accomplished, for example, by profoundly changing the processing path using ODS techniques, for example, which provide a dispersion of fine oxide particles that provide hardening and should also stabilize the microstructure at the dislocation level.

1.3. Austenitic stainless steels

1.3.1. *Introduction*

Stainless steels are extensively used because of their good corrosion and oxidation resistance and because of their high creep resistance. They are thus widely used in the power industry as well in the chemical industry. In these applications, components are often subjected to repeated loading originating from start-up and shut-down cycles. Since the thermal conductivity of these steels is low and the coefficient of thermal expansion is high (compared to ferritic steels) these austenitic steels are subject to thermal fatigue or, quite generally, to creep-fatigue.

These steels all have high chromium concentrations (~17%) that are responsible for the excellent corrosion and oxidation resistance. Additions of austenite formers, such as nickel, stabilize the face-centered-cubic (FCC) structure. The concentration of carbon is intentionally kept very low (~0.01%) in the series that are designated "L" (low interstitial).

The compositions of the four most frequently used stainless steels (SS) are shown in Table 1.2, while their mechanical properties are shown in Table 1.3. In this discussion, there will be very little consideration of the compositions referred to as "stabilized", such as 321SS and 327SS. Most emphasis will be on 316SS that, because of the addition of molybdenum, has better high temperature properties than 304SS. The data in Table 1.3 show that in the annealed state these steels have a rather low yield strength. However, their strain-hardening properties, as indicated by the ratio $S_{uts}/S_{0.2}$, are quite remarkable.

The microstructure of these steels is much simpler than that of ferritic steels, which have been discussed above, and superalloys, which will be discussed below. They are essentially solid solutions with low stacking fault energy (SFE) which explains their high strain-hardening capacity, especially at low temperatures.

At high temperature, these alloys can undergo ageing, which is manifested by the precipitation of chromium carbides ($M_{23}C_6$) at the grain boundaries. Two methods are used to limit this phenomenon. The first is to tie up carbon in the form of highly stable niobium or titanium carbides (used for stabilized steels such as 321SS or 327SS). Clearly this technique is of limited utility in welded structures. The second method is to reduce the level of carbon in the steel. However, since this also decreases the strength, up to about 0.10% nitrogen is added. Nitrogen-bearing steels are denoted as LN.

Material	Heat Treatment	C	Si	Mn	P	S	Ni	Cr	Mo	Ti	Nb
304	1075°C/0.5h	0.07	0.72	1.84	0.027	0.009	8.62	18.29			
316	1050°C	0.07	0.53	1.66	0.023	0.008	10.79	16.75	2.23		
321	1150°C/1h	0.06	0.57	1.10	0.040	0.017	9.00	17.13		0.30	
347	1100°C/1h	0.07	0.57	1.42	0.029	0.014	11.00	18.18			0.79

Table 1.2. *Chemical composition (wt.%) and heat treatment of four stainless steel compositons [YAM 80]*

Material	ASTM Grain Size	Vickers Hardness	Yield Strength (MPa)	Ultimate Tensile Strength (MPa)	Elongation (%)	Reduction in Area (%)
304	3.5	142	92	373	42	73
316	6	149	135	481	44	70
321	3	135	113	353	42	67
347	8	150	155	409	36	63

Table 1.3. *Grain size, hardness and mechanical properties in uniaxial tension at 600°C for four compositions of austenitic stainless steels [YAM 80]*

1.3.2. *Mechanical behavior and microstructure*

These alloys exhibit remarkable cyclic hardening behavior in fatigue, as seen in Figure 1.17. Whereas their elastic limit is lower than the 9–12% Cr steels, they become stronger than the latter when they are cyclically deformed. However, they are significantly weaker than superalloys. This strong cyclic strain-hardening capacity persists when they are subjected to relaxation fatigue, even for long hold times, as seen in Figure 1.17. The strain-softening phenomenon seen in the 9–12% Cr steels is absent in the austenitic stainless steels.

This cyclic strain-hardening capacity is reflected in the cumulative damage diagram in Figure 1.18. We can see here that predeforming by fatigue increases the creep strength. The cumulative damage can be significantly greater than one. For 9–12% Cr steels, diagrams of the same type show that the damage can be significantly less than one [FOU 08]

The dislocation structures resulting from high-temperature plastic deformation in these austenitic stainless steels have not been studied very extensively; certainly not as much as the dislocation structures resulting from room temperature fatigue. However, the results reported in other chapters in [BAT 10] [PIN 10] can be used, as can other studies on this subject [LEV 79, NAH 77, NAH 81]. At high temperatures, cells are formed whose walls are composed of dislocations and whose formation is favored by thermally activated cross slip. Examples of such structures in 316SS are shown in Figure 1.19 [REZ 82]. With a hold time at high temperature, intergranular precipitation of chromium carbides of the $M_{23}C_6$ type is seen that is dependent on the level of carbon present; see Figure 1.20 [REZ 82]. Similarly, cellular precipitation occurs at the triple points of grains in 316SS. This particular form of precipitation gives rise to very pronounced local heterogeneous deformation and grain boundary sliding which, as will be shown later, is the origin of grain boundary cavitation in these steels.

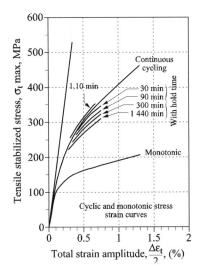

Figure 1.17. *Effect of hold time in tension on cyclic hardening of a 316L stainless steel at 600°C, taken from [MOT 82] results*

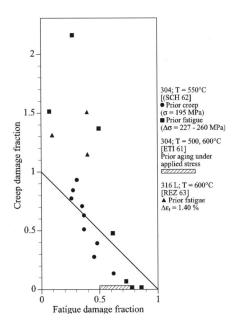

Figure 1.18. *Cumulative damage in creep-fatigue for various austenitic stainless steels*

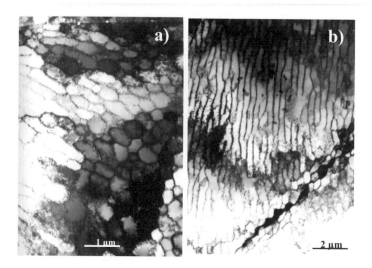

Figure 1.19. *316 stainless steel tested in continuous cycling fatigue at 600°C and at $\Delta\varepsilon_p / 2$ of 0.86%: a) grain containing dislocation cells; b) grain with veins and dislocation walls*

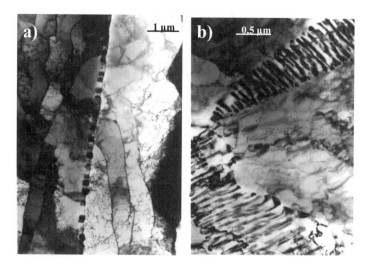

Figure 1.20. *316 stainless steel tested in fatigue-stress relaxation with a hold time (tm) at 600°C: a) $\Delta\varepsilon_t / 2 = 2.3\%$, tm = 0.5h, tf = 35h; b) intergranular precipitation of carbides $\Delta\varepsilon_t / 2 = 0.7\%$, tm = 100 min, tf = 380h. Cellular precipitation of carbides at a triple point [REZ 82]*

1.3.3. *Life and damage*

1.3.3.1. *Continuous cycling fatigue*

Coffin-Manson curves for 316 LN, such as those in Figure 1.21, show that increasing the temperature by 25°C at 600°C leads to a decrease in lifespan of about a factor of three for a given plastic strain range. The effect is even more pronounced when testing is done under alternating torsion.

Results obtained from several different 316 LN steels (SP, SQ, SR, ECL) at several different heats are shown in Figure 1.21. The composition of these various heats of 316 stainless steel are reported in Table 1.4. For more details on the compositions of these, see Argence and Pineau [ARG 96]. These results show a marked difference in behavior between the heats when tested at high temperature. Later it will be seen that this effect is tied to the sensitivity of intergranular fracture in these heats. Figure 1.21 also shows that using a von Mises equivalent strain representation does not normalize the results onto a single curve.

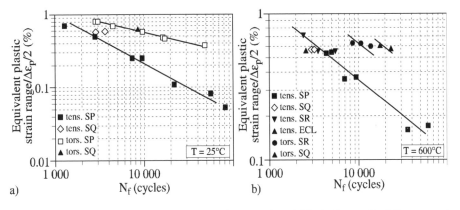

Figure 1.21. *LCF curves for different heats of 316 LN (see Table 1.4) tested in fully-reversed uniaxial fatigue and fully-reversed torsion at: a) 25°C; and b) 600°C [ARG 96]*

The influence of frequency or strain rate on the fatigue life of a 304 SS is reported in Figure 1.22 [YAM 80]. As expected, the lifespan decreases with decreasing cyclic frequency. This reduction is associated with the appearance of intergranular cracking at low strain rates, as shown in Figure 1.23 [YAM 80].

The transition between the two failure modes occurs at higher strain rates with increasing temperature. The appearance of intergranular fracture and the corresponding reduction in lifespan are favored by large grain sizes, as shown by Yamaguchi and Kanazawa's results [YAM 80] in Figure 1.24.

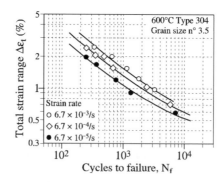

Figure 1.22. *LCF curves for 304 SS tested at 600°C showing the influence of strain rate*

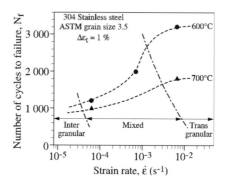

Figure 1.23. *LCF Fatigue life as a function of strain rate for 304 SS tested at 600°C and 700°C [YAM 80]*

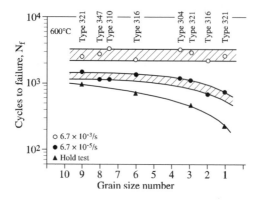

Figure 1.24. *Influence of grain size on the LCF lifespan at 600°C for different types of austenitic stainless steels ($\Delta\varepsilon_t = 1.0$ %). Note that the larger the grain size number, the smaller the grains*

In order to model the fatigue life, it is necessary to consider crack initiation and propagation. Crack initiation can be defined on various size scales. Taking initiation crack size to be in the order of the grain size, typically 20–50 μm for austenitic stainless steels, it is possible to develop a master curve giving the number of cycles to crack initiation, N_i, as a function of the number of cycles to failure, N_f. Such a curve was developed for tension-compression by Jacquelin [JAC 83] and by Levaillant [LEV 84]. Mathematically, it is given by:

$$N_i = N_R - 12N_R^{0.62} + 0.226N_R^{0.90} + 74 \qquad [1.8]$$

The results upon which this equation is based are given in Figure 1.25. A similar expression to equation [1.8] has been put forth in the English design code, R5 [HAL 91]. In compiling results for other stainless steels (e.g. 304) and for other FCC metals and alloys, Hales has shown that there is a "universal law" for all of these materials. This "law" is included in Figure 1.25, where it can be seen that it is equally applicable to metals such as aluminum and nickel.

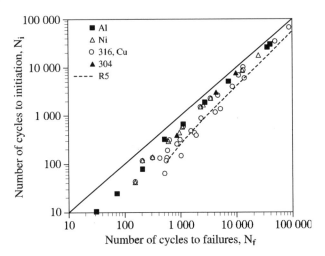

Figure 1.25. *Correlation between the number of cycles to crack initiation and number of cycles to failure obtained for various metallic materials including two types of stainless steels (304 and 316). The curve used in the UK code R5 is also shown*

Such a "law" is useful in modeling the lifespan in tension-compression loading, as will be seen. However, it has the limitation of not being multiaxial and it fails to account for the directional aspect of fatigue crack initiation in fatigue. This aspect can be very significant. The orientation of crack initiation facets at the free surface of a LCF specimen of 316L SS, tested in tension-compression or in alternating torsion, is shown schematically in Figure 1.26 [ARG 96, WEI 93]. At ambient temperature, the initiation facets correspond to Stage I, which is on the maximum shear plane for both uniaxial and torsional loading. At high temperature, the orientation of the initiation facets changes and is normal to the maximum principal stress. This change is associated with the presence of a thin oxide layer that behaves in a brittle manner.

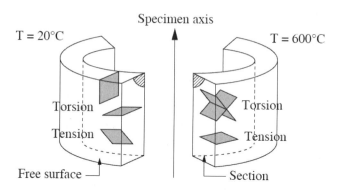

Figure 1.26. *Schematic showing the orientation of crack initiation facets observed at the free surface and on longitudinal cuts. Tests in continuous cycling fatigue at 20°C and at 600°C [ARG 96]*

This change in the orientation of initiation facets is responsible for some rather particular sequencing effects that have been observed in this steel and which are shown in Figure 1.27.

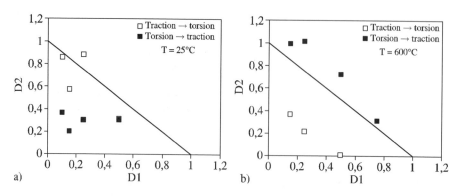

Figure 1.27. *Cumulative damage in fatigue for SQ steel (see Table 1.4, p34). Sequential testing in tension-compression and in torsion at: a) 25°C; and b) 600°C [ARG 96]*

For a more complete discussion, see Weiss and Pineau [WEI 93]. These sequencing tests were carried out as will now be described. The material is first cycled for N_1 cycles in uniaxial tension (or conversely in torsion) to a damage level defined as $D_1 = N_1/N_{F1}$ where $N_1 < N_{F1}$. It is then cycled in torsion (or conversely torsion-tension) for N_2 cycles, which produces a damage level defined as $D_2 = N_2/N_{F2}$, where N_{F1} and N_{F2} refer to the cycles to failure under the corresponding conditions without any sequence effects.

The results reported in Figure 1.27 show a larger departure from Miner's linear damage law ($D_1 + D_2 = 1$).

At ambient temperature, the largest deviation is seen for the torsion-tension/compression sequence, whereas at high temperature just the reverse is seen.

This different behavior at various test temperatures is easily explained, at least in a qualitative fashion, with reference to the directional aspect of the orientation of the crack initiation facets, see Figure 1.26.

At 20°C, the initiation facets in torsion can propagate easily in tension/compression since some of them are suitably oriented normal to the loading axis. However this is not the case for crack initiation facets formed first in tension, which must turn in order to propagate in the torsion mode.

At high temperature, the inverse of this behavior is seen. For example, the facets initiated in torsion are no longer suitably oriented to propagate in tension. Instead, the crack formed in torsion at high temperature must turn to propagate in tension.

It may also be noted that at room temperature cracks form in response to the maximum shear stress, characteristic of dislocation movement. At high temperatures, the cracks initiate in response to a maximum dilatation. This may be associated with an environmental effect, since ingress of a foreign species – in this case oxygen – would be enhanced by a dilatational component of stress.

In order to exploit these results in a quantitative manner, it would be necessary to know the initiation and propagation "laws" for both loading modes in order to build a more complete model [WEI 93].

1.3.3.2. *Hold time effects*

The effects of a hold time in tension for tension-relaxation fatigue on the high temperature fatigue life of austenitic stainless steels have been the object of numerous studies. Since this effect depends upon the composition of the steel in question, detailed work carried out at Ecole des Mines de Paris on five different compositions will be relied upon in the following discussion. The compositions of these steels are given in Table 1.4.

Stainless	C	N	Ni	Cr	Mo	Mn	S	P	B
VIRGO	0.033	0.025	13.60	16.40	2.12	1.55	0.0220	0.022	0.0012
SP	0.021	0.080	12.27	17.18	2.40	1.74	0.0080	0.033	0.0032
SQ	0.028	0.077	12.46	17.31	2.44	1.88	0.0010	0.028	0.0012
SR	0.024	0.072	12.10	17.40	2.33	1.70	0.0030	0.018	0.0010
ELC	0.011	0.068	11.07	16.64	2.14	0.85	0.0057	0.028	ND

Table 1.4. *Composition of five different austenitic stainless steel heats, all nominally 316 LN, that were used in relaxation fatigue studies*

The results obtained for a 316 L SS (VIRGO) tested with long hold times are shown in Figure 1.28.

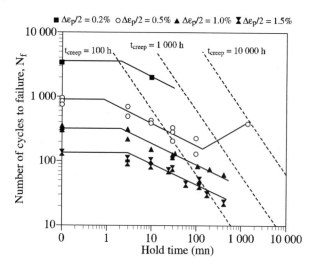

Figure 1.28. *VIRGO steel (Table 1.4). Influence of hold time in tension on life. Tests of relaxation fatigue in tension-compression were carried out at various levels of plastic deformation at 600°C. The isochronal curves giving the rupture times t_f as a function of the hold time and number of fatigue cycles are also shown [ARG 96]*

Isochronal curves for rupture are also given in this figure. It can be seen that for short hold times (several minutes) the number of cycles to failure for a given level of strain varies only slightly, whereas at longer hold times the number of cycles to fracture decreases regardless of the level of strain applied. It must be emphasized here that the time to fracture increases, which shows that failure does not only depend on time. A test of very long duration (about one year) with one cycle every 24 hours shows that the number of cycles tends to increase. Similar behavior has been reported for other heats of 316L SS (see, e.g. [LEV 82, LEV 88, WAR 81]. This saturation of the hold time influence on lifespan is less apparent when the steels are tested at lower temperatures (550°C). Pre-ageing likewise plays a role on the saturation effect. For a complete discussion, see Levaillant *et al.* [LEV 81].

The reduction in lifespan that is seen with increased hold time is closely related to the appearance of intergranular fracture in all of the specimens. Some examples are shown in Figure 1.29.

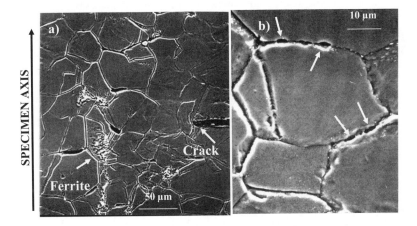

Figure 1.29. *Examples of intergranular damage in austenitic stainless steels tested in creep-fatigue: a) SR steel (Table 1.4) tested in tension/compression with $\Delta\varepsilon_{t\,eq}\,/\,2 = 1\%$, and $t_h = 0.5h$. Optical microscopy of a longitudinal section. Note the presence of cavities in the body of the sample, far from the main crack which is indicated with an arrow; b) steel SR tested in tension/compression having the same conditions as in a). Nucleation of intergranular cavities is shown by arrows [ARG 96] and [WEI 92]*

It can be seen that intergranular damage is not only around the major crack but appears to be uniformly distributed throughout the sample. This observation indicates that at least during the crack propagation phase, intergranular fracture in creep-fatigue of austenitic stainless steels is not related to environmental effects, contrary to that which is seen in superalloys. In the case of stainless steels, intergranular damage is due to creep, which occurs during hold times, and not only just in front of the crack but also in the bulk of the specimen.

1.3.4. *Physically-based damage modeling of creep-fatigue interactions*

In this section two physically-based models will be presented, the goal of which is to take hold time effects into account. The first model relies upon grain boundary sliding. Since few data are available for this type of slip in stainless steels, this model must be considered to be phenomenological. The second model relies directly upon measurements of intergranular damage. It was developed in an attempt to evaluate the residual life of components based on measurement of the degree of intergranular cavitation.

1.3.4.1. *Grain boundary sliding*

The literature on creep is replete with examples that show the importance of grain boundary sliding on the formation of intergranular cavities (see e.g. [MCL 62]). On the other hand, very little information is available on this subject in fatigue (see e.g. [PIN 83]). McLean and Pineau [McL 78] nevertheless have attempted to use the measurements of grain boundary sliding in creep carried out by Gates [GAT 77] to explain the effects of hold time in tension reported by Udoguchi *et al.* [UDO 73]. Several assumptions must be made in order to pursue this concept. In particular, the notion of an effective level of grain boundary sliding, S, was defined as follows:

$$S = S_p^t + S_p^c + 0.30 S_h \qquad [1.9]$$

where S_p^t and S_p^c are the components of grain boundary sliding during the loading and unloading parts of the cycle respectively and S_h is the grain boundary sliding that occurs during the hold time. Results are provided in Figure 1.30, which shows a good correlation between the value of S calculated using relationships given by Gates [GAT 77] and the number of cycles to failure. This correlation gives satisfactory results at 650°C and 750°C. Note that the slope of the two straight lines in Figure 1.30 is -0.40, and not -1.0 which would imply that the total grain boundary sliding calculated to the point of fracture is not constant but depends on the test conditions, in particular on the amount of strain and temperature.

1.3.4.2. *Extensive intergranular damage.*

Since the 1980s development of a correlation between the degree of intergranular damage and creep-fatigue life has been sought (see e.g. [LEV 84]). Intergranular damage has been defined as:

$$D = \frac{L_f}{L_t} \qquad [1.10]$$

where L_f is the cumulative length of cavitated grain boundaries per unit area and L_t is the total grain boundary length per unit area. These measurements are made on a two-dimensional photograph. L_f is measured in the body of the specimen, far from any crack. Examination of broken samples from creep-fatigue tests has led to the observation that, to a first approximation, D varies linearly with the number of cycles (see e.g. Hales [HAL 80]). The damage per cycle can thus be defined as $D_c = D/N$.

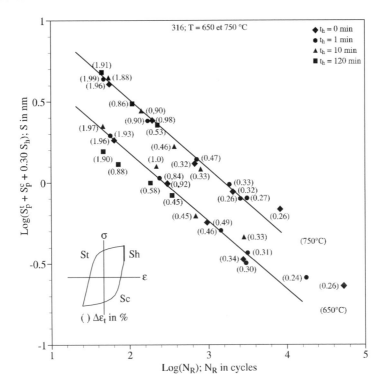

Figure 1.30. *Correlation between calculated intergranular grain boundary sliding (S) and the number of cycles to failure (N_R) of 316 SS [MCL 78]*

As a start, D_c, which is specific to the test conditions (temperature, strain, hold time) has been experimentally determined.

The results obtained for four types of 316L SS are shown in Figure 1.31.

Significant steel-to-steel variations are seen. In particular, the VIRGO steel, with the least resistance to creep, is the one that exhibits the highest values of D_c, all other factors being equal. As expected, D_c increases with increasing hold time. It should be noted that for the VIRGO steel, D_c decreases for the longest hold times (1,440 minutes), which is in good agreement with the increase in the corresponding lifespan, see Figure 1.28. This observation lends credence to using D_c to represent creep-fatigue damage.

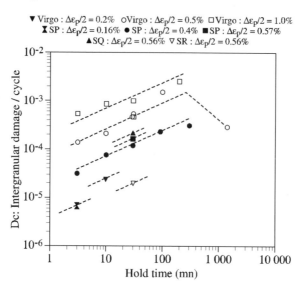

Figure 1.31. *Influence of hold time on intergranular damage per cycle, D_c. Test results from various compositions of 316L SS (see Table 1.4) [ARG 96]*

Measurements of D_c served as a basis for the development of a crack propagation model in creep fatigue [LEV 82, LEV 88, PIN 82, PIN 83a, PIN 83b]. It is assumed that the application of a hold time introduces oxidation and intergranular damage such that, to a first approximation, crack initiation occurs at $N_i^{CF} = 0$, where N_i^{CF} is the number of cycles to crack initiation in creep-fatigue. More over, it is postulated that a crack acceleration relationship for intergranular damage has the following form:

$$\left(\frac{da}{dN}\right)^{CF} = \left(\frac{da}{dN}\right)^{PF} \frac{1}{\left[1 - k\, D_c^{p-1}\, D\right]^2} = \left(\frac{da}{dN}\right)^{PF} \frac{1}{\left[1 - k\, D_c^{p}\, N\right]^2} \qquad [1.11]$$

where $(da/dN)^{CF}$ and $(da/dN)^{PF}$ are the rates of crack propagation in creep-fatigue and in pure fatigue respectively. Parameter k must be determined, as must exponent p. It should be noted that Skelton proposed a rather similar expression starting from the notion of accumulation of linear damage [SKE 93]. We will return to Skelton's approach later.

Integration of equation [1.11] leads to a simple expression giving the relative reduction of propagation life in creep-fatigue, N_p^{CF}, compared to that in continuous fatigue, N_p^{PF}:

$$R = \frac{N_p^{PF} - N_R^{CF}}{N_p^{PF} \times N_R^{CF}} = k D_c^p \qquad [1.12]$$

where N_R^{CF} is the lifespan in creep fatigue cycling.

The correlation established between D_c and the reduction in lifespan, R, is shown in Figure 1.32 for two steels of the 316L family. It is observed that the results are grouped around the same straight line, which allows for the determination of exponent p ($p \sim 0.923$).

Hales' results [HAL 80] are also shown in Figure 1.32. This author used another definition for intergranular damage, which no doubt explains the shift in the two sets of results. The results corresponding to the longest hold times (1,440 minutes) are well-situated on the curve, in spite of the fact that an increase in lifespan is seen for the longest hold times (see Figure 1.28).

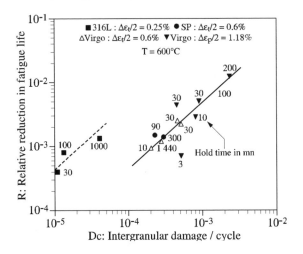

Figure 1.32. *Correlation between the intergranular damage per cycle, D_c, and the relative reduction in lifespan at 600°C for two different compositions of austenitic stainless steels [ARG 96]*

If the damage calculation methodologies just discussed are to be used in a truly predictive manner, it would be necessary to calculate the intergranular damage D_c

(or D) starting from the relaxation curves. It is for this reason that an intergranular damage evolution expression due to creep was established by studying the creep of notched specimens [YOS 86, YOS 90]. An incremental multiaxial expression was developed having the form:

$$dD = B\Sigma^{\gamma}\left(\varepsilon_{eq\,vp}\right)^{\delta}d\varepsilon_{eq\,vp} \tag{1.13}$$

where B varies only with temperature, γ is a constant $(2<\gamma<4)$ and $\delta=0.5$.

The integration of equation [1.13] with $\gamma=4$, along the entire stress relaxation curve yields:

$$D_c \approx \frac{2B}{\sqrt{E}}\left(\sigma_{t\,max}\right)^{4}\left(\sigma_R\right)^{1/2} \tag{1.14}$$

where E is Young's modulus, $\sigma_{t\,max}$ is the maximum value of the stress at the start of relaxation and σ_R is the relaxed stress [LEV 88]. This expression was applied to three variants of 316L and good results were obtained, as shown in Figure 1.33.

Based on the above discussion, a complete provisional schematic methodology, such as that shown in Figure 1.34, may be suggested. It must be noted that this approach, validated for the moment at only one temperature (600°C), is worth extending into other temperature regimes and should be all the more useful for longer hold times. This supposition stems from the fact that creep-fatigue initiation was assumed to be instantaneous.

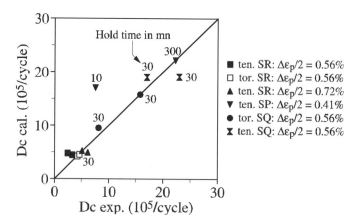

Figure 1.33. *Comparison between calculated and measured per cycle intergranular damage values, D_c, for three variants of 316L (see Table 1.4) [ARG 96]*

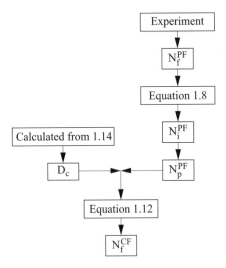

Figure 1.34. *Schematic methodology for lifespan prediction of austenitic stainless steels under creep-fatigue loading and assuming intergranular damage*

1.3.4.3. *Metallurgical implications*

Intergranular precipitation of $M_{23}C_6$ in austenitic stainless steels has been extensively studied (see e.g. [WEI 72]). This precipitation depends upon both temperature and carbon level in the steel and is accelerated by cyclic deformation [WEI 93]. This carbide is sometimes formed by a cellular phase transformation, as was illustrated in Figure 1.20. Among the steels shown in Table 1.4, the variants denoted as VIRGO, SR and SQ are those that show the most sensitivity to hold time. They are also those in which a cellular phase transformation was seen.

It is equally well-known that a certain number of minor elements, notably boron, exert a strong influence on intergranular precipitation and grain boundary cohesive strength. This could explain, at least partly, the superior properties exhibited by the SP variant, which has a higher boron concentration (32 ppm). It is known, in effect, that this element has a strong influence on the creep ductility of these steels [BOU 77] and, as a consequence, their creep-fatigue life. Thus, at least in part, small differences in composition can explain the significant variations in creep-fatigue life. As has been shown, these compositional modifications are manifested in terms of the degree of intergranular damage that is observed; see Figure 1.31.

Intergranular precipitation does not always have negative effects, as long-term testing of the VIRGO variant has shown (see Figure 1.28). This saturation effect, even when beneficial, at long hold times could be due to a reduction in grain boundary sliding brought about by intergranular precipitation. As will be seen later in this chapter, the legacy of cast Ni-base superalloys have utilized grain boundary carbides for decades to inhibit grain boundary sliding and creep damage. In any event, it is clear that more work needs to be done to explain these ageing effects on the long-term behavior of austenitic stainless steels.

Intergranular damage, being all the more significant when the degree of relaxation is important (see e.g. equation [1.14]), there is interest in developing steels that do not show significant relaxation during hold periods, that is materials that have a good creep life. This certainly explains the superiority of the 316 LN variants compared to 304 SS, which do not contain molybdenum. This conclusion relating to mechanical behavior is the opposite of what will be shown in the section on Ni-base superalloys. It is striking how fundamentally different these two classes of materials are in terms of their primary high temperature damage mechanisms and behavior.

1.4. Fatigue of superalloys

1.4.1. *Microstructure and processing of superalloys*

1.4.1.1. *Phases and microstructure*

As mentioned previously, superalloys (both Ni-base and Ni-Fe base) are used in the hot sections of jet engines and, although less frequently, also in the power generation industry. Microstructurally, the Ni-based alloys consist primarily of small coherent precipitates embedded in a FCC or γ matrix. The precipitates, denoted γ′, have an L1$_2$ crystal structure based on the Ni$_3$Al-ordered compound, as seen in Figure 1.35.

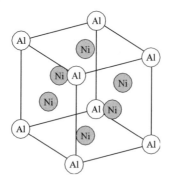

Figure 1.35. *Crystal structure of Ni$_3$Al γ′. In the idealized structure, there are three total atoms at the centers of the cube faces and one total atom at the cube corners. FCC-like crystallographic structure is designated L1$_2$*

While microstructurally simple, they are somewhat chemically complex. In addition to Al, they have Ti (which is substituted for Al in the γ', increases the fraction of γ', and decreases the fault energy in γ'), Cr (primarily in γ, confers oxidation resistance), and Mo (in both γ and γ' may increase or decrease the relative mismatch depending on other elements). Some residual carbon may be present in alloys used for disks (relatively low temperature) as an impurity due to processing (carbon boil), while in other alloys it may be intentionally added to form carbides that reduce creep by inhibiting grain boundary sliding. Other elements may be added to provide solid solution strengthening (e.g. Ta) to control the volume fraction of γ', to inhibit grain boundary sliding by migrating to boundaries (e.g. odd sized atomic species, such as Zr) to improve castability (Hf) and to control the shapes of carbides. The volume fraction of γ' in these alloys ranges from 25% to over 50%, with more modern alloys having a higher volume fraction. In addition, the mismatch between the precipitate and matrix is important in deformation and in precipitate stability. The mismatch is defined quantitatively as:

$$\delta = \frac{a_{\gamma'} - a_{\gamma}}{\overline{a}} \quad\quad\quad [1.15]$$

where the numerator represents the differences in lattice parameter and the denominator represents the average lattice parameter. Typically, mismatch parameters range from almost zero (e.g. Waspaloy) to as high as 0.4%. The γ' precipitates may be spherical (low mismatch) or cuboidal (high mismatch). The effect of various elements on superalloys has been widely studied and much detailed information may be found in [SIM 72, SUP 08, TIE 89]. Over the years, as the function of various elements has become better understood, compositions have been simplified. To summarize, the following major phases are present in Ni-base alloys:

– an FCC γ matrix;

– coherent precipitates having an LI_2 crystal structure and denoted by γ';

– carbides, depending on the alloy and intended application.

These alloys are processed differently, depending on the intended application. The major forms in which superalloys are used are:

– single crystals (for jet engine turbine blades, with a $[001]$ axis aligned along the blade axis);

– DS, which are polycrystalline but with all crystals aligned longitudinally in the $[001]$ direction (for land-based turbines);

– billets formed from forged or rolled ingots (disks);

– powder metallurgy products (HIP (Hot Isostatic Pressing) or HIP+Forged, used in jet engine disks).

Similar considerations apply to the Ni-Fe base superalloys, such as In 718, that in addition to Fe has Nb added to form coherent precipitates with a disk-like structure [PAU 69] in a FCC γ matrix. These precipitates have a DO$_{22}$ crystal structure and are based on the Ni$_3$Nb ordered compound, shown in Figure 1.36, and are denoted by γ′. As with the γ′ Ni-base alloys, alloying elements such as Al and Ti may be added and occupy the Nb sites.

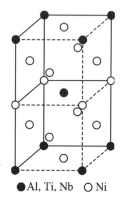

● Al, Ti, Nb   ○ Ni

Figure 1.36. *Crystal structure of γ′. This crystal structure is denoted DO$_{22}$ and may be viewed as stacking two FCC-like ordered unit cells to produce the complete unit cell*

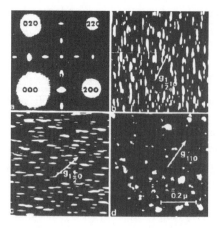

Figure 1.37. *The γ′ morphology in In 718 is revealed by dark field TEM and shows a disk-like phase with the long axis along different <100> directions. In (d) the disks are in the plane of the page [PAU 69]*

The disk-like shape of the γ′precipitate is shown in Figure 1.37 via dark-field electron microscopy that was used to selectively image different variants of γ″. In addition, there is some γ′ present on which the γ″ has been shown to precipitate. These alloys are primarily used in disks in jet engines as well as components in the power generation industry. Components from these alloys are generally produced by traditional ingot metallurgy followed by mechanical forming to obtain a uniform fine grain size (ASTM 6–7). In summary, the microstructure of these alloys consists of: (i) an FCC γ matrix; (ii) coherent precipitates having a DO_{22} crystal structure and denoted by γ″ with some γ′; and (iii) a fine, generally uniform, grain structure.

Representative compositions are given in Table 1.5.

1.4.1.2. *Processing of superalloys*

While the main topic of this section is fatigue of superalloys, it would be remiss not to say a few words about processing, especially since processing greatly affects the final properties and is closely linked with the product form.

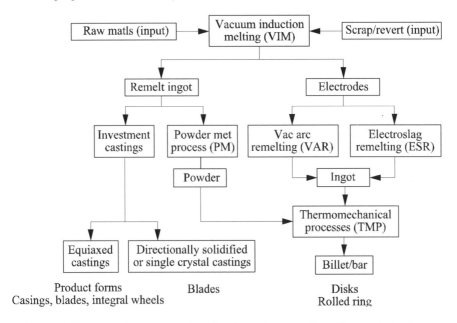

Figure 1.38. *Processing routes and product forms for superalloys. Vacuum technology, introduced in the 1950s, has led to high-purity materials with low inclusion content*

Product Form/Alloy	Ni	Fe	Co	Cr	Mo	W	Ta	Nb	Al	Ti	V	C	B	Zr	Hf	Y	Re
									Element (wt. %)								
CAST/WROUGHT																	
IN-718	50	30		18	3			5.2	0.6	0.8		0.05	0.006				
Waspaloy	58		13	19	4.3				1.3	3		0.08	0.006	0.06			
INVESTMENT CAST																	
Rene 77	58		15	15	4.2				4.3	3.3		0.07	0.015	0.04			
Rene 80	60		9.5	14	4	4			3	5		0.17		0.03	0.8		
POWDER METALLURGY																	
Rene 95	Bal		8.1	12.8	3.6	3.6		3.6	3.6	2.6		0.08	0.01	0.053			
N18	Bal		15.5	11.5	6.5				4.3	4.3		0.02	0.015		0.5		
IN-100	Bal		18.5	12.4	3.2				5	4.3	0.8	0.07	0.02	0.06			
Udimet 720	Bal		14.7	16	3	1.25			2	5		0.025	0.02	0.03			
CAST SINGLE CRYSTAL																	
CMSX - 2	66		4.6	8	0.6	7.9	5.8		5.6	0.9		0.005					
CMSX-4	Bal		9	6	0.6	6	7		5.6	1					0.1		3
Rene N4	63		7.5	9.2	1.6	6	4	0.5	3.77	4.25							
PWA 1480	63		5	10		4	12		5	1.5							
PWA 1493*	Bal		4.9	9		3.8	5		3.6	1		0.02	0.03	0.07		0.02	

* Mo+Ta+W = or > 10%

Table 1.5. *Some representative compositions of superalloys (various sources)*

Since these materials are used in rotating machinery and nuclear power plants, safety is a major concern and safety is directly related to the cleanliness of the material, among other factors. Use of vacuum melting technology was pioneered in the superalloy industry and is used to reduce or eliminate undesirable and reactive elements that can show up as undesired compounds or constituents in the final product. These alloys are often produced by vacuum induction melting (VIM) followed by vacuum arc remelting (VAR) and the electro slag remelting (ESR) to produce high cleanliness alloys, of which In 718 is an outstanding example. Various processing routes and product forms are shown in Figure 1.38.

1.4.2. *Deformation mechanisms*

In this section, a brief review of basic deformation mechanisms is provided since an important aspect of fatigue damage at any temperature is related to the way in which deformation occurs. Superalloys provide an excellent vehicle for this discussion. They contain most of the features of interest and have the advantage of being microstructurally simple. They have been extensively studied.

In most engineering alloys, we may make the following observations:

– Deformation at low temperatures is generally by precipitate shearing since the precipitates are closely spaced and coherent. At higher temperatures, deformation-induced coarsening may occur, leading to dislocation looping and lower strength.

– For the shearing mode of deformation, strength generally increases with precipitate size, while for looping the strength decreases. This has been observed experimentally and predicted theoretically. Maximum strength occurs at the shearing-to-looping transition. Dislocation shearing is shown in Figure 1.39 for Waspaloy, a Ni-base superalloy widely used as a disk material in jet engines.

– At elevated temperatures and low strain rates, the rate-limiting step for monotonic deformation is a creep-like climb process. If the material is polycrystalline, grain boundary sliding is an additional mode of deformation.

– The fundamental mode of deformation may change as a function of temperature. For example, cross slip from octahedral {111} to cube {100} planes occurs with increasing ease in the γ' precipitates in Ni-base alloys as the temperature is increased. However, the mobility of dislocations on {100} planes is very low (high Peierls force) and the cross slipped segments act as barriers to dislocation glide on the octahedral planes, thus increasing the hardness of γ' with increasing temperatures. In a γ / γ' alloy this gives rise to a regime in which the yield strength is relatively insensitive to temperature.

– Due the ease of cross slip and the basic γ′ structure, there is an observed asymmetry in the tension and compression yield stress, especially in single crystals, which may have profound effects on crack formation and propagation, and hence on the life of such materials.

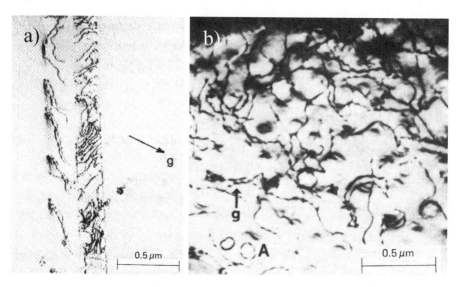

Figure 1.39. *Observations of dislocations in Waspaloy via TEM using a [200] diffraction vector. a) Fatigue cycled at 25 and* $\Delta\varepsilon_p = 0.06\%$. *Dislocation pairs are typical of precipitate shearing in Ni-base superalloys; b) same conditions as for a) except that the material was treated to have larger precipitates. Dislocations looping the precipitates are seen (e.g. region A on micrograph) [LER 84]*

1.4.3. *Cyclic deformation and microstructure*

1.4.3.1. *Low and intermediate temperatures*

Since Ni-base and Fe-Ni base alloys are used in jet engine disks, where both fatigue crack propagation and LCF are important design parameters, some discussion of cyclic deformation at these temperatures is appropriate. Cyclic deformation of Ni-base alloys has been reviewed extensively elsewhere [ANT 89] and only the salient points will be reviewed here.

During fatigue in a strain-controlled test, the stress responds in accordance to the applied strain levels and the influence of the microstructure. When an annealed specimen undergoes strain-control fatigue, the stress generally increases with increasing cycling for a certain number of cycles. Depending on the deformation mode and its internal stress state, the response stress will either stabilize or reach a maximum stress followed by softening. The response stress is also dependent upon the test temperature, strain rate and the applied plastic strain amplitude, since these variables strongly influence the deformation substructure.

1.4.3.1.1. Cumulative glide and cyclic stress strain behavior

Numerous superalloys harden upon cyclic straining both at room temperature [CLA 82, FOU 77, HUR 86, LER 84, LER 85a, MER 74, STO 78, WEL 64, WEL 67] and at elevated temperatures [BRE 83, COF 71, LER 84, MER 74, MIN 82 WEL 65, WEL 67]. This hardening is a result of an increase in the dislocation density and extensive cross-slip at higher temperatures, all of which restrict the motion of mobile dislocations and constitute deformation debris. Theories to explain hardening then involve the accumulation of forest dislocations or the formation of Cottrell-Lomer locks. However, at lower temperatures the dislocations are restricted to the primary {111} slip planes, as cross-slip is limited. Hardening at lower temperatures, where planar slip predominates, has been explained in terms of a competition between precipitate shearing and increased slip band density [LER 87].

Basically this theory explains the relatively low cyclic hardening at low strains in terms of precipitate shearing on single slip bands since relatively few active bands are required to carry the imposed strain. At higher strains the hardening becomes much more pronounced, since there are multiple slip-bands and dislocation debris at slip intersections. During cycling, more slip-bands must nucleate because of the restricted slip distances and eventually an equilibrium density of slip bands is achieved. In essence, the distance between slip-band intersections acts as a kind of barrier akin to grains in polycrystalline single-phase materials.

1.4.3.1.2. Applications to Waspaloy and Nimonic 80A

Waspaloy and Nimonic 80A are two superalloys that contain about 20–25 v/o of the strengthening γ' phase. They exhibit planar, inhomogeneous slip at low and intermediate temperatures with precipitate shearing by paired dislocations [CLA 80, CLA 82, DIE 84, LER 82, LER 84, LER 85a, MER 74, STO 78], as seen previously in Figure 1.39a for Waspaloy. As discussed above, slip-band density increases with strain, and follows a kind of Hall-Petch relationship, as seen in Figure 1.40 for Nimonic 80A [LER 83] and Waspaloy [LER 84].

Mathematically this relationship may be expressed as:

$$\sigma = \sigma_o + ki^{-1/2} \hspace{4cm} [1.16]$$

where σ is the maximum stress in a cumulative glide plot, i is the slip band spacing, and σ_o, k are constants for a given material.

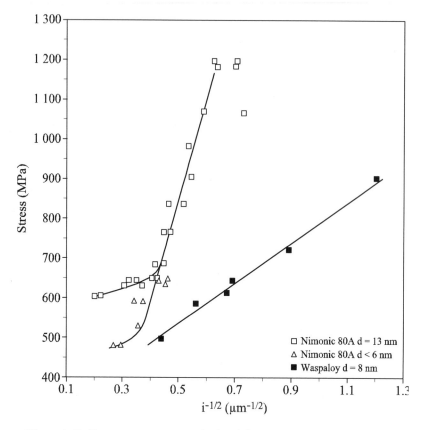

Figure 1.40. *Stress response* versus *slip band density in Nimonic 80A [LER 87] and Waspaloy. The precipitate sizes are indicated on the graph. The cyclic strain hardening exponent is indicated by n' [LER 84]*

This equation is valid for plastic strain ranges that induce multiple slip. At low plastic strain ranges, most grains deformed elastically or by dislocation motion on a single slip system and hardening was minimal. As expected, these two deformation mechanisms and damage accumulation modes gave rise to different strain hardening exponents, n', as seen in Figure 1.41. Similar observations have been made in other superalloys after fatiguing at 650°C [MIN 82].

Eventually, an equilibrium number of slip bands forms and hardening ends. The response stress then decreases in numerous superalloys [FOU 77, LER 84, LER 85a, MER 74, STO 78, WEL 64, WEL 67] and the material cyclically softens by extreme precipitate shearing as has been observed in In 718 [FOU 77, MER 74], Waspaloy [CLA 82, LER 84, MER 74, STO 78] and Nimonic 80A [LER 85a]. Figure 1.42a shows extreme shearing in cyclically deformed Nimonic 80A [LER 85a] with a bimodal γ′ distribution.

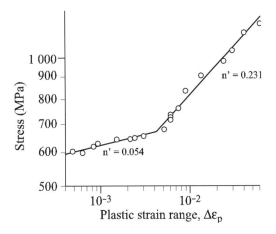

Figure 1.41. *Cyclic stress-strain curve of Nimonic 80A at 25°C showing bi-linear strain hardening behavior [LER 83]*

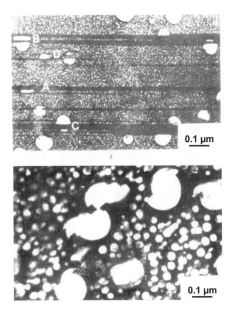

Figure 1.42. *Nimonic 80A [LER 85] a) and Waspaloy [CLA 80]; b) showing extreme shearing of the γ′ precipitates. Shearing is easily seen in the large precipitates, but the small ones control the slip mode*

The shearing process occurs through dislocation motion along {111} planes. These planes often occur in pairs separated by a distance of 10–50 nm [LER 85]. The material between these pairs is displaced in one direction. To achieve this, a large shear in one plane must be compensated for by a shear in the other plane that is equal to but opposite in magnitude that of the first. This results in an irreversible shear on one plane but a completely reversible unit when a pair of planes is considered. Thus, reversed macro-shearing is carried out by irreversible micro-shearing on shear-coupled planes. This can be observed in Figure 1.42a for Nimonic 80A and in Figure 1.42b for Waspaloy. Careful TEM work shows that there is a small tilt in these regions that takes the precipitates out of Bragg conditions and makes it appear as if they were dissolved. A small compensating tilt re-images the precipitates and shows that the precipitates are not resolutioned, as suggested by Merrick [MER 74]. Resolutioning thus does note explain the softening behavior in these alloys.

In the absence of precipitate shearing, the stress saturates. This effect was seen in studies designed to investigate deformation mode and fatigue at constant strength levels. In these studies, heat treatments were carried out in which the γ' precipitates were small (i.e. 5–8 nm diameter) or large (duplex, 300 +90 nm diameter) and the strength was constant. Saturation behavior was observed in Waspaloy [LER 84], [STO 78] containing the large γ' particles. This structure promoted dislocation looping and stress saturation. With continued cycling, the number of loops left around each particle increased until the particles were hidden from view by a tangle of dislocations, see Figure 1.43. In fact, a point was reached where annihilation and creation were occurring at the same rates and the structure was in dynamic balance. Although some slip bands were observed in this alloy, the dislocation structure was much more homogeneous than that observed for the material containing the small shearable particles.

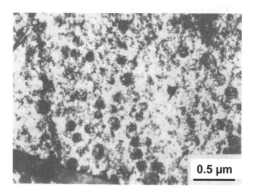

Figure 1.43. *Dislocation looping in Waspaloy at 25°C. The black circles result from a high density of loops left around the γ' precipitates [LER 82]*

The material having a more homogeneous dislocation structure work hardened more rapidly [STO 78], resulting in a higher stress response for a given strain range. This differs from the small precipitate material where the increase in stress due to the higher dislocation density can be partially offset by the softening component due to shearing. This relaxation mechanism cannot occur in the material containing large precipitates and the stress continues to increase until the applied strain and deformation substructure have been fully equilibrated.

The homogeneous deformation retards crack initiation but a crack, once formed, will propagate more rapidly because of the higher stress. On the other hand, when shearable precipitates are present, slip concentration leads to the formation of ledges at the surface and early initiation. However, the softening component tends to slow down crack propagation. Recall that the yield stress for both planar and homogeneous slip materials was identical [LER 84, STO 78].

Thus, any differences in the response stresses or fatigue life are a direct consequence of the differences in deformation mechanisms and damage accumulation modes. Similar behavior was observed at elevated temperatures for several superalloys [MIN 82] in which the addition of very large γ' particles promoted a more homogeneous dislocation structure.

Other deformation mechanisms have been observed during strain cycling of In 718. This alloy has been investigated at both room and elevated temperatures [FOU 77].

At room temperature, it was found that the dislocations moved in thin bands parallel to {111} planes. Thin micro-twins were also observed and were suggested to be a result of the shearing process. Dislocation debris and sheared precipitates were the result of complex deformation mechanisms required by the DO_{22} structure.

The complex deformation mechanism also leads to the formation of orthorhombic δ delta (sometimes called β in the technical literature). This phase transformation along with the shearing-induced reduction in the mean precipitate diameter within the slip plane led to cyclic softening [FOU 77].

Sanders *et al.* [SAN 81] also observed that fatigue deformation occurred by micro-twinning at lower temperatures in In 718.

1.4.3.2. *High temperature deformation mechanisms*

As the temperature is increased, the thermal activation enhances cross-slip in these materials. Examples of this can be found in Waspaloy and Nimonic 80A. As the temperature increases, the dislocation structure becomes more homogeneous. At

800°C, slip bands in Nimonic 80A [LER 83a, LER 85b] are still observable, but are not well defined; see Figure 1.44.

The formation of planar slip bands indicates that shearing still occurs even at 800°C. It should be noted that the density of dislocations within the bands was much lower than at lower temperatures. This suggests that the applied plastic strain can be accommodated by fewer dislocations, since they are more mobile at this temperature. Softening was observed at this temperature from the beginning of cycling. The number of slip bands per grain was very small with very few slip band intersections. Further, the number of dislocations within these bands was small. Therefore, the dislocations were relatively free to move within the grain. According to the hardening model presented earlier, no hardening should occur since slip-band intersections are absent, which is consistent with the observed results. Shearing of the γ', however, still occurred. This resulted in softening. This softening behavior at elevated temperatures was also observed in Waspaloy [LER 82, LER 84]. Similar softening was seen in René 80 [ANT 81b], an alloy in which the volume fraction of γ' and mismatch parameter are both higher than in either Waspaloy or Nimonic 80A.

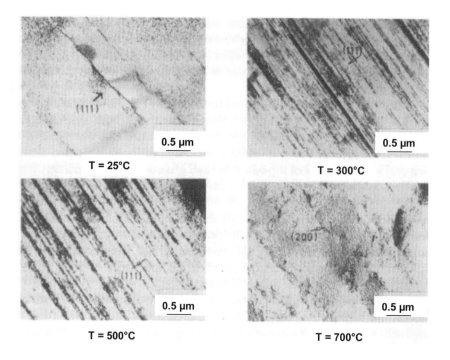

Figure 1.44. *Slip-band density as a function of temperature for Nimonic 80A tested at a plastic strain range of 0.15% [LER 83a]*

Increases in temperature led to microstructural changes and hence to deformation mechanisms that are not normally observed at ambient temperatures. In several superalloys [BRE 81, BRE 83, COF 71, LER 82, LER 83, LER 84, WEL 67], a strain ageing effect occurred during cycling in the intermediate temperature regime. In Waspaloy [LER 82, LER 84], specimens tested with both shearable and non-shearable precipitates hardened during the test when cycled at 500°C. TEM investigations revealed a very high dislocation density within each grain. The number of slip bands in each grain was correspondingly high. Similar observations were made by Bressers and Verhegghe in Waspaloy [BRE 81] and also in Nimonic 80A at 300°C and 500°C [LER 83].

The effect of temperature on the slip-band density is shown in Figure 1.44 for Nimonic 80A. The number of slip bands and the high dislocation densities restrict the motion of mobile dislocations, increasing the response stress. These specimens had higher response stresses than were observed for specimens tested at equivalent plastic strain ranges at any other temperature.

It has been suggested [LER 84, WEL 67] that in these temperature ranges the mobile dislocations become impeded and the formation of additional dislocations or slip bands is required to accommodate the applied strain. This hardening has been explained [HAY 82] as a Portevin-Le Chatelier effect in Waspaloy at 538°C involving a carbon atmosphere around dislocations. However, calculations have been made [KOU 82] showing the mobility of carbon atoms is too high to exert a significant drag force on the dislocations. The strain ageing effect has therefore been attributed by several authors [BLO 85, CHU 76, KOU 82, LLO 75] to other less mobile solutes. However, this is also the temperature range at which the γ' particles will precipitate and coarsen, which could also cause hardening. It has also been found that carbides precipitate on slip bands at 700°C [CLA 82, LER 84]. Precipitate on dislocations cause the dislocations to become immobile and the nucleation of fresh mobile dislocations is required to carry the imposed plastic strain range. Similarly, coarsening of the γ' will hinder the motion of some dislocations, requiring an increased stress for continued deformation. Also, in this temperature range, cross-slip onto {100} planes has been observed in pure γ' [LAL 79, MIN 82]. This process results in an increase in the yield stress of the single crystal PWA 1480 with temperature up to about 800°C [SHA 84]. A similar argument might be applied to Waspaloy and the other alloys to explain this hardening effect at intermediate temperatures.

As the temperature increases, the slip mode becomes more homogeneous and alloys deforming more homogeneously should harden to a greater extent than those deforming through planar glide, as seen in several Ni-base superalloys tested at 650°C [MIN 82]. This was not the case for either René 77 or René 80. In René 77 [ANT 81a] fatigued at 927°C, a stable dislocation structure developed early in the lifespan. Most dislocations were stored at the γ / γ' interface and were shown to be near-edge in character. A similar network of near-edge dislocations was observed in René 80 at 871°C and 981°C [ANT 79a, ANT 81b]. These interfacial dislocations reduced the misfit energy of the system and as such did not constitute damage. They did, however, accommodate the growth of the precipitates. This was consistent with the continual stress drop with increasing cycling, as observed in the cumulative glide plot. The large precipitates give a larger mean free path for dislocation motion and, assuming deformation took place by looping the response stress, would be expected to decrease.

Up to this point, we have tacitly assumed that when shearing took place, it did so by the passage of super partials of the type a/2 <110>. Now it has been shown that at high temperatures dislocations propagate by the motion of a/6 <112> dislocations in the γ phase (see e.g. Milligan [MIL 87]). Since this mode of deformation involves diffusion, it can only take place at low deformation rates, especially at intermediate temperatures. This deformation mode provides a basis for understanding the monotonic and cyclic deformation behavior of single crystalline PWA 1480 that was studied as a function of temperature in these loading modes [MIL 87].

At low temperatures, deformation occurred by shearing of the γ' particles. At intermediate temperatures, a transition occurred between shearing and a by-pass mechanism. The temperature at which this transition occurred was dependent upon the strain rate. At high strain rates the deformation proceeded similarly to that observed at lower temperatures, i.e. shearing of the γ' by a/2 <110> dislocations. At lower strain rates, shearing occurred through the passage of a/6 <112> partial dislocations, which led to the formation of intrinsic stacking faults in the precipitates. At still higher temperatures, dislocation loops were observed in the matrix and at the γ / γ' interface. An accumulation of near-edge dislocations at the interface was also observed. These were shown to lie on {101} planes having a/2 <110> Bürgers vectors.

1.4.3.3. *Summary of cyclic deformation mechanisms*

Cyclic hardening results through an increase in the dislocation/slip-band density at low temperatures. Interactions between intersecting slip bands impede the further motion of dislocations increasing the internal stress.

The reduction of the precipitate diameter within the slip plane due to shearing of the γ' results in cyclic softening. This is a consequence of the inhomogeneous nature of slip in these alloys.

Large precipitates are looped by dislocations resulting in cyclic stress saturation. The high constant stress associated with such materials reduces the fatigue life as compared to small precipitate materials, which soften.

At elevated temperatures, the increased mobility of the dislocations leads to extensive cross slipping. More homogeneous dislocation structures including networks at the γ/γ' interface are observed.

At intermediate temperatures, a dynamic strain ageing mechanism often occurs that increases the response stress and reduces both the ductility and the fatigue life. There is also a corresponding increase in the dislocation and slip band density.

At higher temperatures and low strain rates, the precipitates coarsen. The degree of coarsening depends on the γ/γ' mismatch and the diffusion rates of the alloying elements.

In γ'-strengthened systems, micro-twinning is an important deformation mechanism.

1.4.4. *High-temperature low-cycle fatigue*

LCF is a life-limiting process that occurs in turbine blades where temperatures range from about 760°C at the platform to more than 1,200°C near the blade tips, depending on the alloy. It is also an important process for disks, especially near the rim in the region of the attachment points.

While most modern turbine blades are in the form of single crystals aligned along the [001] axis, many engines are still in service that use polycrystalline cast alloys, such as René 80. These alloys are characterized by large grains, grain-boundary carbides to restrict grain-boundary sliding, a γ matrix, and a high volume-fraction of γ' precipitates, with a mismatch parameter of about 0.3%.

In this section, high-temperature LCF will be considered for both polycrystalline and single crystalline superalloys.

1.4.4.1. *High-temperature LCF of cast polycrystalline superalloys*

René 80 is a legacy Ni-base alloy that has been (and is) widely used for turbine blades. It is a good example of high-temperature LCF behavior. Extensive studies have been carried out at different temperatures, strain rates and hold times [ANT 81a]. LCF test results are shown in Figures 1.45 and 1.46, and cumulative glide behavior is shown in Figure 1.47.

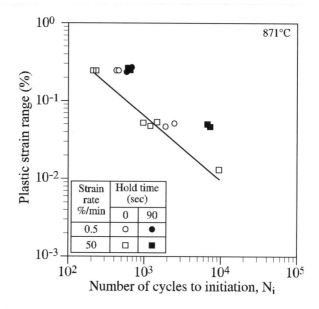

Figure 1.45. *LCF behavior of René 80 at 871°C for the conditions indicated on the graph [ANT 81b]*

It is important to note that when presented on the basis of plastic strain, the LCF life at 871°C is not reduced by tensile hold time and in fact may even be increased. There is continuous softening throughout the duration of the test.

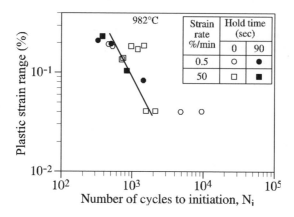

Figure 1.46. *LCF behavior of René 80 at 982°C for the conditions indicated on the graph [ANT 81b]*

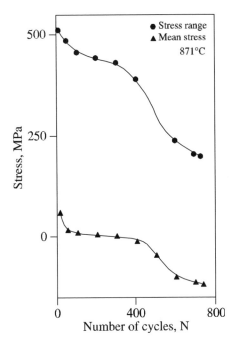

Figure 1.47. *Stress range and mean stress for LCF of René 80 at 871°C cycled at 0.5% / min showing continuous softening [ANT 81b]*

These observations differ from those usually made that show that the lifespan is reduced by creep damage, as previously seen in the section on 9–12% Cr steels and austenitic stainless steels. Likewise, they also differ from the numerous cases where initial rapid cyclic hardening is followed by stress saturation for the majority of the lifespan, as was seen in the austenitic stainless steels. On the other hand, this continuous cyclic softening during deformation is seen in the 9–12% Cr steels. Clearly these results indicate that lifespan predictions based on strain range partitioning (SRP) [MAN 71, SAL 88] or frequency-modified fatigue life [COF 72] approaches require some modification. Microscopic optical and TEM studies provided a basis for understanding this behavior. It was determined that:

– the initial precipitate structure significantly coarsened and small precipitates were consumed by large ones (Figure 1.48);

– oxidation along boundaries was frequently observed (Figure 1.49a);

– internal cracking of grains normal to the stress axis was also seen (Figure 1.49b).

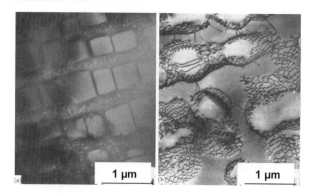

Figure 1.48. *Dark field TEM micrograph: a) showing the initial microstructure of René 80; and b) microstructure after LCF testing at 981°C. Note the bimodal distribution of in a) and the interfacial dislocations in b) which have been shown to be near edge in character. The small precipitates have been consumed by the large ones during LCF testing at $\dot{\varepsilon} = 0.46\% / min$ and $\Delta\varepsilon_p = 0.196\%$ [ANT 81b]*

Similar observations were made for René 77 [ANT 81b] in which it was shown that coarsening required plastic deformation and such plastic deformation greatly accelerated coarsening. Simply holding René 77 at a high temperature without any stress had negligible effects on the microstructure.

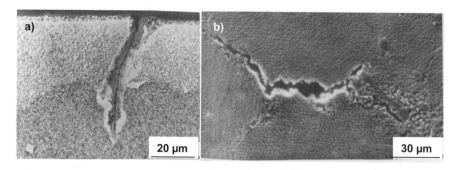

Figure 1.49. *Oxide spike: a) for testing at 871°C, $\dot{\varepsilon} = 0.50\%/min$, $\Delta\varepsilon_p = 0.25\%$; and b) boundary cracking normal to load axis, $\dot{\varepsilon} = 0.50\%/min$, $\Delta\varepsilon_p = 50\% / min$ with 90 s at the maximum strain [ANT 81b]*

The fatigue life of a specimen or a component may then be viewed as a trade-off between the stress (which may be greatly relaxed in a strain-controlled test by coarsening) and the formation of oxide spikes or the embrittlement of boundaries by oxygen. The low dislocation density observed in fatigue at high temperatures does not support the notion of deformation debris being very damaging. In fact, the majority of dislocations were observed at the γ /γ' interface and were edges in character to accommodate the mismatch between these two phases. In fact, these dislocations would actually reduce the strain energy in the system and may be viewed as geometrically necessary.

These observations and hypotheses were used to develop a criterion for LCF lifespan prediction in which cracking is viewed as the result of a competition between oxide formation and the response stress, which in a strain-controlled test is controlled by precipitate coarsening.

The formation of oxides and oxide ridges has long been recognized as an important mechanism of damage in Ni-based superalloys (e.g. see [COF 74]). The goal of the following work was to incorporate oxidation in a specific and physically consistent manner in the development of a quantitative formulation. Basically, increased coarsening results in a lower stress, which means that a longer oxide spike must be present before crack initiation. Of course, a longer oxide means that more cycles must accumulate.

A quasi-fracture mechanics criterion based on metallographic observation was used for crack initiation and parabolic oxidation was invoked based on other studies [PED 82]. Using these assumptions, it was possible to show that the cycles to initiation, N_i, are given by:

$$N_i = C_2 \, x \left(\frac{v}{1+vt_h} \right) \exp\frac{Q}{RT} \cdot \Delta\varepsilon_p^{-8n'} \qquad [1.17]$$

where C_2 is a material constant, v is the cycle frequency, t_h is the hold time (if any), n' is the strain-hardening exponent, Q is the activation energy for diffusion of the damaging species (generally oxygen), T is the absolute temperature and R is the gas constant.

This equation has the form of Coffin's frequency modified law with an explicit evaluation of the frequency dependence as well as the hold time. This equation has been put into parametric form and successfully used to predict the lifespan of turbine blades.

The notion of the lifespan being determined by a trade-off between coarsening and a reduced stress and oxygen diffusion is further strengthened by experiments on René 80 in air that were carried out over a large temperature range [BOU 81]. The results of these tests are shown in Figure 1.50.

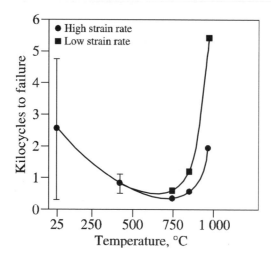

Figure 1.50. *Low cycle fatigue of René 80 tested in fatigue at constant strain range at high and low strain rates over a large temperature range. Note the generally longer lives at lower strain rates and the minimum lifespan at 760°C [BOU 81]*

These results, obtained at a constant strain range, show that the lifespan decreases with increasing temperature to about 760°C, at which point there is a minimum. Above that temperature, the lifespan, for a constant plastic strain range, actually increases. These results may be understood in terms of the tradeoff mentioned above.

At temperatures of up to about 760°C, the environmental effect dominates this tradeoff. The structure is stable, the stresses are high and the environment is increasingly damaging. However, above this temperature the precipitates coarsen very rapidly and the response stress decreases. Thus the degrading effect of the environment, although more severe on an absolute basis, is more easily tolerated and the lifespan again increases with temperature. It should be emphasized that this behavior is expected to depend on the lattice mismatch, δ, as discussed above.

Part of the driving force for coarsening (in addition to the size differences) is tied to the effect of stress on the equilibrium shape of the precipitate under stress, as discussed elsewhere [PIN 76]. In general terms, the higher δ, the greater the thermodynamic driving force for microstructural change. This change can be accommodated by diffusional mass transport. Thus more rapid transitions are expected for high-mismatch alloys (e.g. PWA 1480 and René N4) than for low-mismatch ones (e.g. Waspaloy and Udimet 500).

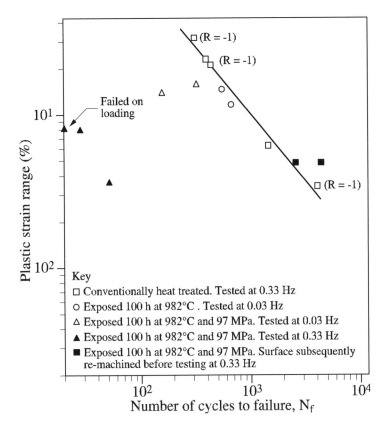

Figure 1.51. *Effect of prior exposure under stress on the LCF behavior of René 80 at 871°C. All testing was done at $R_\varepsilon = 0.05$ [ANT 79a]*

Additional studies were carried out to examine the effects of long-term exposure for samples that were stress-free and those under a very small stress (about one-third of the yield stress) [ANT 79a]. The results of those studies are summarized in Figure 1.51, in which conventional LCF test results are compared to the LCF behavior for samples that were pre-exposed. In addition to precipitate coarsening, the exposure also allowed oxidation to occur.

Both pre-exposure conditions reduced the fatigue life to below that for a conventional test (heat-treated, machined and tested). However, stressed exposure reduced the life by one to two orders of magnitude. Stress-exposed samples exhibited surface-connected boundary damage that was a combination of oxidation and formation of $M_{23}C_6$ boundary carbides; see Figure 1.52.

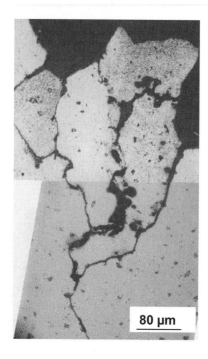

Figure 1.52. *Fracture surface region of a pre-exposed René 80 sample showing extensive grain boundary damage. Tested at 871°C, cyclic frequency of 0.33Hz, $R_\varepsilon = 0.05$, and a total strain range of 0.57% [ANT 79a]*

The lifespan was restored and even enhanced beyond the conventional one when the surface-connected damage region was removed by machining. These results clearly demonstrate that conventional LCF test results are non-conservative and that life prediction needs to take physical mechanisms into account.

Studies on a Co-base superalloy have been carried out and reported in detail elsewhere [REU 83a, REU 83b, REM 87]. The composition (wt.%) of the alloy that was used was 0.59C, 11Ni, 23.2Cr, 6.95W, 3.32Ta, 0.37Zr, 0.22Ti, 0.17Fe, 0.008B, 0.005P, 0.003S, and the balance was cobalt. These alloys are widely used in the cast form as vanes in jet aircraft and, although not used in rotating components, are subjected to thermally-induced loading as well as air-flow loading. They have a microstructure that includes a high fraction of carbides in the boundaries and matrix, which undergo oxidation in use, especially along the grain boundaries. This work demonstrated that in high-temperature fatigue, oxidation is a primary damage mode. Extensive measurements were made that demonstrated that the oxidation rate depended upon the inelastic strain and that the depth of grain boundary carbide oxidation was proportional to $t^{1/4}$. A model was proposed in which damage was assumed to be a linear combination of the fatigue crack-growth rate without oxidation and the oxidation rate (for both the matrix and grain boundary). While this formulation may seem to exclude interactions, such is not the case because, as was clearly demonstrated, the carbide oxidation rate increased with inelastic strain. Tomkins' model for plastic crack propagation was used [TOM 75].

Based on these detailed experimental observations, the equation for this model was:

$$\frac{da}{dN} = 0.51\Delta\varepsilon_{in}\left[1/\cos\left(\frac{\pi\sigma}{2T}\right)-1\right]a + (1-f_c^*)\alpha_M^o\left(1+K_M\frac{\Delta\varepsilon_p}{2}\right)(\Delta t)^{\frac{1}{2}} +$$

$$f_c^*\alpha_C^o\exp\left(b\frac{\Delta\sigma}{2}\right)(\Delta t)^{\frac{1}{4}} \qquad\qquad [1.18]$$

where the first term is related to pure fatigue, the second term is related to the matrix component of cracking in the matrix due to oxidation (seen to depend on inelastic stain) and the third term relates to carbide oxidation along the boundaries (also seen to depend upon the inelastic strain through the stress). In this equation f_c^* refers to the fraction of carbides in the crack path. The αs are quasi-oxygen diffusion constants for the matrix and carbide and depend on temperature through an Arrhenius relationship. The cycle time is denoted by Δt. This model can be integrated in a straightforward way by noting that the FCP component of cracking for both matrix and boundary oxidation are independent of crack size and essentially constant. Thus equation [1.18] has the form:

$$\frac{da}{dN} = Ba + C \qquad\qquad [1.19]$$

where C is the sum of the oxidation terms, $(da/dN)_{0x}$, and B is the crack length multiplier in equation [1.18]. This equation may be integrated quite easily to give the cyclic life N_f:

$$N_f = \frac{1}{B}\ln\left[1+\frac{Ba_f}{\left(\frac{da}{dN}\right)_{ox}}\right] \qquad\qquad [1.20]$$

This equation has been used with some success in predicting the lifespan of Co-base alloys, at least for well-behaved cycles.

1.4.4.2. *Low- and high-cycle fatigue of single crystalline materials*

Single crystal turbine blades have replaced those fabricated from cast polycrystalline materials, especially in the higher performance military and civilian aircraft. The single crystal turbine blades have greatly extended the lifespan by eliminating grain-boundary sliding as a creep damage mode and by reducing thermally-induced stresses, since they are oriented with the long axis of the blade in the low modulus $[001]$ direction. They are generally produced by a grain selection method [DUH 89, GEL 86], in which one favorably oriented grain is allowed to grow through a choke. While this ensures the $[001]$ axis is along the blade vertical axis, the secondary orientation is generally not controlled. This means that the envelope of the blade is essentially randomly oriented with respect to the crystal axes and it is known that the secondary axis has a large effect on fatigue properties, as shown in the section on fatigue crack propagation. This uncontrolled variable has the potential to exert "scatter" in the LCF lifespan of blades made from this material.

Additional challenges include anisotropy, which is a major issue in analysis and lifespan prediction. In addition, the precipitate morphology may change in service from cubes to plates [PIN 76, REE 99] (normal or parallel to load axis) and tension/compression asymmetry may occur [SHA 84]. These latter two effects depend upon the relative orientation of the load axis and the crystalline axes. Such factors may have a profound effect on the creep and fatigue properties [REE 99]. In the following two sections, the high- and low-cycle fatigue (both design-critical properties) of these materials are considered.

1.4.4.2.1. LCF of Ni-base single crystals

As has been mentioned, LCF is a design-critical property, both in and of itself. More important is the analysis of the lifespan around notches. Notches occur at the attachment points of the blade to the disk and at the small cooling holes in the blade. They raise the stress and strain concentrations (and may be analyzed by using techniques previously developed in [PIN 10]) and are areas in which fatigue cracking is most likely to occur by a micro LCF process. It is thus important to understand the LCF behavior in single crystalline materials.

Some widely used Ni-base single crystals are PWA 1480/1493, CMSX-4 and René N4. An extensive study of LCF of Ni-base single crystals was carried out on the single crystalline alloys PWA 1480/1493 and model alloy SC 7-14-6 (Ni-6.8% 13.8% Al and 6% Mo), which are representative of the aforementioned materials [ARA 01]. LCF tests were carried out over a range from 25–982°C at relatively high plastic strains in air and in H_2 gas, which is representative of the space shuttle main engine. Specimens were tested in different orientations under strain control at R-ratios close to zero (zero-tension-zero cycling), except for a few cases of large negative and positive values. Testing was carried out at 10 Hz. The lifespan at 649°C is plotted in terms of plastic strain range in Figure 1.53, where it can be seen that there is considerable "scatter". However, under these conditions the fracture surface was crystallographic and it would seem natural that fracture should be controlled by some combination of shear and tensile parameters (strain range, stress range).

The notion is that shear stresses/strains are necessary to cause dislocation motion which is turn causes "damage", and tension across the fracture plane is necessary to open up the crack. The various theories have already been discussed in a chapter in [PIN 10] and will not be repeated here in detail.

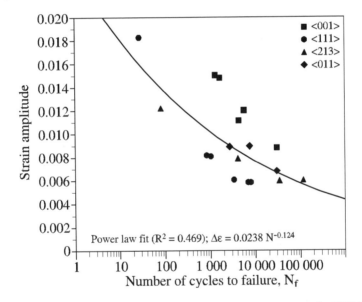

Figure 1.53. *Cycles to failure versus longitudinal plastic strain range for PWA 1493 at 649°C and at various orientations [ARA 01]*

Calculations of stress and strain quantities on the octahedral and cube slip planes in directions corresponding to slip directions were made and then substituted into various multiaxial cracking criteria. An example of one such correlation is shown in Figure 1.54. For a more detailed examination of different criteria for multiaxial fatigue failure, we can refer to Chapter 1 of [BAT 11].

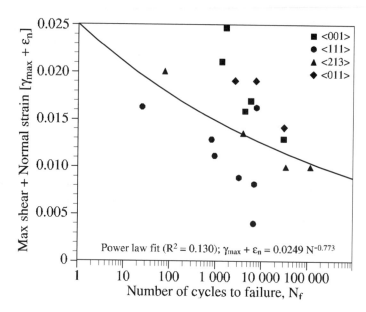

Figure 1.54. *Graph of* $[\gamma_{\max} + \varepsilon_n]$ *versus* N_f *for PWA 1493 at 649°C and at various orientations [ARA 01]*

In this graph the shear strain is the strain on a critical plane and the tensile strain is the strain across the plane on which the maximum shear occurs [KAN 82]. As can be seen, the correlation is unacceptable. More complex criteria involving shear and normal components were invoked, but no improvement in "scatter" was seen. The first two presentations of the data (as well as others not shown here) may lead us to believe that what is being observed is the essential variability of this material. However, these alloys are highly controlled in terms of composition and processing parameters. Defects such as inclusions and low angle boundaries are minimized. It is thus highly unlikely that the scatter is real. It is most likely that the representation of the data is inappropriate. When the data were plotted in terms of the maximum shear stress range, the data for all orientations tended to show significantly less "scatter", as shown in Figure 1.55.

This improved correlation tends to indicate that the principal factor in controlling lifespan is the development of damage through the movement of shear-driven dislocation motion, intersection and precipitate destruction. A more physics-based model could be developed to further improve the data correlation. As mentioned previously, stress is not a true damage driver but rather a mathematical "stand in" for strain. Additional tests at higher temperatures or in a hydrogen environment at 24°C, showed far less crystallographic fracture and correspondingly a much poorer correlation. In this regime, damage initiated in the γ/γ' eutectic regions along the interphase regions, which are not crystallographic. When tested at a higher temperature in H_2, the crystallographic effect returned, since it is well-known that high temperatures diminish the ability of H_2 to stay trapped within the material.

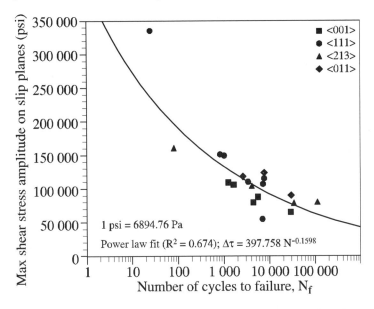

Figure 1.55. *Graph of τ_{max} versus N_f for PWA 1493 at 649°C and at various orientations [ARA 01]*

1.4.4.2.2. High-cycle fatigue of Ni-base single crystals

Blades in the turbine region are cooled using "cool" compressor air that sets up mean stresses that may vary in time and with stress amplitude, as shown schematically in Figure 1.56 for three hypothetical loading paths [WRI 04]. These paths illustrate what can happen during actual flight and are meant to show time dependence. For example, consider the start-up phase of an engine. Flame impingement on the blade at a given location introduces a compressive component. Of course, the centrifugal force will impose a relatively constant tensile mean stress. With time, the compression will relax and there will be a drift towards a higher tensile mean stress associated with the centrifugally-induced stress.

These blades are also subject to vibratory stresses that may be induced by sufficiently large inclusions. Since they are envisioned for long-term use, they are coated with protective layers to minimize the effects of the environment.

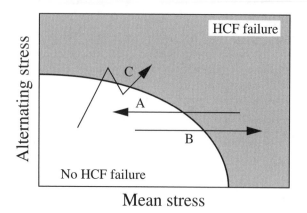

Figure 1.56. *Modified Goodman diagram showing possible loading paths during high-cycle fatigue (HCF) of a turbine blade. The arrows show possible loading paths during operation [WRI 04]*

An extensive study of high-temperature, high-cycle fatigue of the single crystal alloy PWA 1484 was carried out for a range of frequencies from 6–900 Hz and R-ratios from -1 to +1 at 1,038°C, with the goal of identifying the type of damage and then using that information to build an engineering model [WRI 04]. It was found that at low R-ratios (<0) and high frequencies the damage was essentially cycle-dependent and independent of frequency, which indicates failure by a classical fatigue mechanism. However, at higher R-ratios and lower frequencies, fracture was dominated by a creep rupture mechanism. It was also found that there was damage mode mixture over a very significant portion of the Goodman diagram, with creep rupture constituting 50% of the damage at R = 0.1. These experimental findings were then used to develop models of varying levels of sophistication. The fatigue lifespan was modeled using a modified Wöhler approach:

$$N_f = K_f \sigma_{eq}^n \tag{1.21}$$

where N_f is the cycles to failure, K_f is a constant for the material and conditions of interest, n is fitting constant and σ_{eq} is the equivalent stress taking into account the R-ratio. The equivalent stress is defined through the following equation:

$$\sigma_{eq} = \sigma_{alt}\left(1 - R^{w-1}\right) \tag{1.22}$$

where σ_{alt} is the alternating stress component and w is an adjustable constant. The creep component was modeled according to the relationship:

$$t_r = K_r\left(\overline{\sigma}\right)^n \qquad [1.23]$$

where t_r is the rupture time, K_r is a constant for the material and conditions, $\overline{\sigma}$ is the effective creep stress and n is an experimentally-determined constant. An extensive data set was available to determine the various required constants.

The effective stress may be thought of as an appropriately defined "average" stress over the cycle. Another view is that stress should be nonlinearly weighted to take into account the nonlinear effect of stress on creep rate. All of these techniques were investigated and it was found that the best agreement was obtained when the effective creep stress was simply taken to be the mean stress in a cycle. Damage was calculated using a simple linear damage rule, as is discussed below. That equation is the well-known "time and cycle fraction" rule given by:

$$\frac{N}{N_f} + \frac{t}{t_r} = 1 \qquad [1.24]$$

The predicted results are shown in Figure 1.57, in which the effective creep stress was taken to be the mean stress in a cycle.

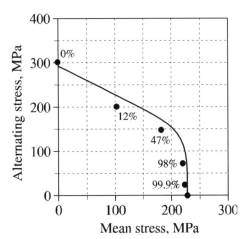

Figure 1.57. *Stress amplitude versus mean stress for 10^7 cycles of high-cycle fatigue life of PWA 1484 at 1,038°C. The solid line represents the predicted lifespan based on equation [1.24]. The points represent averages of experiments. Testing was carried out at 59 Hz [WRI 04]*

This is an example of an engineering study of lifespan prediction for Ni-base superalloys which, while not of a fundamental nature, does recognize the different damage modes present and uses empirical equations to build the model based on actual physical processes, in this case on the basis of creep and fatigue. We may surmise that environmental effects were of less importance due to the effectiveness of the coating used to exclude the environment.

1.4.4.3. *LCF of powder metallurgy alloys*

Powder metallurgy alloys are used to make disks in jet engines because of savings in materials, processing and machining time and because powder products are essentially isotropic. René 95 is an alloy that has been used in powder form in the hip+forged condition for disks in jet engines as has N18, which is discussed below. It is a high-strength alloy with relatively low toughness. Studies were carried out on this alloy [ANT 79b] to assess the effects of hold time on LCF lifespan since disks are subjected to long hold times during flight. Testing was carried out in fully reversed strain in strain control ($R_\varepsilon = -1$), at 649°C and a cyclic frequency of 0.33 Hz. In some experiments, 900 second hold times were applied at the maximum load. This study showed that:

– the effect of an extended hold time was to extend the lifespan when represented in terms of inelastic strain;

– in some cases cracking initiated at pre-existing pores due to incomplete consolidation, even in the forged material;

– for initiation at equivalent pores, the lifespan was an order of magnitude greater when the initiation was subsurface;

– regardless of the initiation mechanism (pores, slip bands, inclusions), there was always a small region of transgranular propagation followed by intergranular propagation.

These results clearly demonstrated the huge effect of the environment on fatigue life, especially in the early stages of crack propagation. The transition from transgranular to intergranular further strengthened this conclusion. Finally, the absence of a hold time effect when representing the data in terms of inelastic strain demonstrates that creep damage, at least for these test conditions, is of secondary importance. Indeed, given the fact that the lifespan actually *increased* with hold time, it may reasonably be conjectured that some relaxation around the crack tips was occurring, giving rise to blunting, slower crack-growth rates and longer lives.

The effects of environment will be discussed in a more rigorous manner in a subsequent section (1.4.5).

1.4.4.4. *Other metallurgical instabilities at high temperature*

In section 1.1.4.1 it was shown that dynamic changes in the material structure have major effects on fatigue life and cannot be ignored. In addition to the changes in precipitates, other dynamic changes also occur. In a study to show such effects using Waspaloy as a model material, it was shown that carbides precipitate on slip bands, see Figure 1.58, which also impinge upon oxidized boundaries to nucleate microcracks, see Figure 1.59.

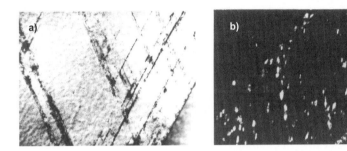

Figure 1.58. *Bright field (a) and dark field (b) TEM micrographs of Waspaloy LCF tested at 650°C. The image in (b) was taken using a carbide diffraction spot. Comparing (a) and (b) shows that carbides precipitated on slip bands and the grain boundaries [LER 84]*

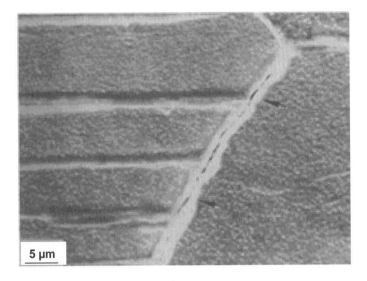

Figure 1.59. *Slip-band-induced cracking of oxidized grain boundaries in Waspaloy tested in LCF at 527°C. Microcracks are indicated by the dark arrows [LER 84]*

Carbides basically lock the dislocations in the slip bands, thereby blocking relaxation of the high stresses at the tip of the slip band. These stresses in turn crack the oxidized grain boundaries, causing crack initiation.

These experiments, and much of the preceding discussion, demonstrate that instabilities (or more precisely dynamic changes) occur:

– at surfaces (embrittlement in the surface region, deformation-induced changes in the surface roughness);

– at the level of the microstructure (precipitate coarsening, dynamic precipitation on slip bands); and

– at the level of the dislocation substructure (slip-band dispersion, annealing of dislocations).

The inescapable conclusion is that generalized approaches to lifespan prediction are unlikely to be successful. Instead, the critical damage mechanisms must be incorporated into material/test condition-specific models. In essence, a table in which test conditions and materials are the rows and columns, and the squares are filled in with specific proved models that take into account the mechanisms. At this point it must be emphasized that we in no way suggest using fundamental physics models for engineering lifespan prediction of real-world components. Instead, we advocate (indeed, insist on) using such models to provide guidance as to the functional form, with parameters being determined experimentally.

1.4.4.5. Thermomechanical fatigue in Ni-base superalloys

Up to this point fatigue has been discussed in isothermal terms and during steady-state operation, which is an appropriate stress state. There are, however, many situations in which both the temperature and imposed mechanical strain vary simultaneously in a repetitive manner. This is termed thermomechanical fatigue (TMF). For example, in very simplified terms the start-up and shut-down phases of a jet engine constitute a TMF cycle. Also alterations in power during altitude changes or other maneuvers of a jet aircraft introduce TMF cycles because of the changes in power that cause fluctuations in temperature as well as changes in loading. An analytical approach to TMF which has been considered earlier in the introduction of this volume, an overview is given elsewhere [ANT 02] and physically-based modeling of TMF is discussed in detail in [MIL 93]. A very detailed analysis of TMF can be found in [CHA 09].

In terms of testing, temperature and strain change simultaneously. The changes in strain arise from a combination of thermal and mechanical effects. Since the thermal strains are not associated with loading, the thermal strain is subtracted from the total strain to give the mechanical strain. The mechanical strain is the control parameter in a TMF test. We may define various paths in terms of temperature and strain. The most revealing kinds of tests are either in-phase or out-of-phase in terms of temperature and strain. The out-of-phase test, which is generally the most damaging, is a test in which the minimum strain is at the maximum temperature.

A schematic of the stress/strain response in a so-called bi-thermal test is given in Figure 1.60 in which the total (mechanical plus thermal) strain is shown for purposes of clarity. This test is meant to provide a kind of bridge between the simple isothermal LCF test and the complex fully-TMF test in which a plethora of possible mechanisms and interactions may be encountered in each cycle. In the bi-thermal test, the interactions are between damage mechanisms at high and low temperature that are thought to capture the most damaging interactions. It has already been shown that superalloys are particularly susceptible to environmental interactions (see e.g. Figure 1.59) and form slip bands, especially at low temperatures (see e.g. Figure 1.44). It is easy to imagine that the oxides that form at high temperatures are easily cracked by the slip bands that form at the lower temperatures and that the out-of-phase cycle could be quite damaging. On the other hand, if the test is of the in-phase type then the stresses at the lower temperature are reduced, the slip-band interaction should not be as damaging and the lifespan would be expected to be longer. That this is in fact true is shown in Figure 1.61, where the in-phase and out-of-phase test results are compared to isothermal LCF tests. It can be seen that the out-of-phase test is the most damaging, especially at low strains and for long lives, as would be expected based on physical reasoning.

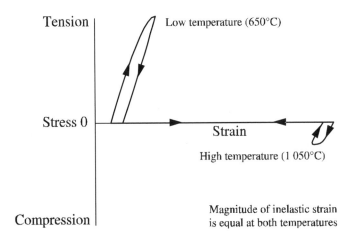

Figure 1.60. *Illustration of out-of-phase TMF cycle for PWA 1480. Cycling was carried out between 650°C and 1,050°C. The straining directions are shown by the arrows. For an in-phase cycle, the maximum strain would at the maximum temperature and the path would be qualitatively reversed [GAY 88]*

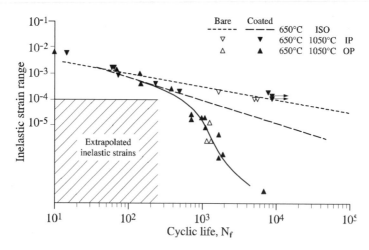

Figure 1.61. *Fatigue lifespan curves for PWA 1480 in both in-phase (IP) and out-of-phase (OP) testing compared to isothermal fatigue at 650°C. Data are provided for bare and coated specimens [GAY 88]*

A further factor to consider is the fact that these materials are generally coated when used as turbine blades in a jet engine. Basically the same kind of considerations apply to the bare and coated specimens. In addition, parts of the coating may spall off, exposing bare metal to engine operating conditions. The coating has a larger coefficient of thermal expansion than the underlying alloy and at the lower temperature in the cycle would therefore be subjected to an increased tensile strain, which would be expected to cause early cracking. In fact, coating cracks do form at the lower temperature and propagate directly into the underlying metal.

This simplifying approach of the bi-thermal test was anticipated by an earlier study [BOU 81]. In these experiments, a René 80 specimen was cycled first at 760°C to about half of its life (~ 270 cycles) at an imposed strain range of 0.05%. It was then cycled at this strain range at 25°C. This is called a hi-lo cycle. The specimen broke on the first loading cycle at 25°C, even though according to isothermal fatigue data the sample would have been expected to last for many hundreds of cycles. Microscopic examination showed that in the hi-lo cycle, slip bands at 25°C caused cracking of the oxides that were formed at 760°C. When the loading sequence was reversed (lo-hi cycle) with no change in the loading parameters, fracture occurred at the "normal" number of cycles at the higher temperature for the imposed loading conditions. The damage mechanisms at RT (room temperature) did not interact with the high-temperature mechanism, which was oxidation related. Furthermore, when the experiment was repeated by cycling for about one-quarter of the lifespan at high temperature, the room temperature lifespan was over 6,000 cycles. This suggests that the damage accumulation rate is decidedly nonlinear as well as being sequence dependent. The same kind of sequence dependence based on damage mechanism was discussed earlier for 316 SS.

1.4.5. *Fatigue crack propagation (FCP) of superalloys*

1.4.5.1. *Introductory comments*

One of the major applications of superalloys is in the disks of jet engines (both Ni-base and Fe-Ni base) where they are used in the high-pressure compressor and turbine sections of the engine. In these applications, the temperatures vary from the bore of the disk to the rim, depending on the function and location in the engine. Typical temperatures may be as low as 350°C at the bore to as high as 700°C at the fir-tree where the turbine blades are attached. Important design parameters are LCF and FCP.

An example of the importance of FCP in jet engines was seen in the crash of United Airlines Flight 232, which was flying as scheduled between Denver and Philadelphia via Chicago. On July 19, 1989, the Douglas DC-10 suffered an uncontained failure of its number 2 engine, which destroyed all three of the aircraft's hydraulic systems, freezing the control surfaces. The failure initiated in a Ti fan disk containing an elliptical crack at the bore whose surface trace was about 12 mm. Failure analysis demonstrated that the crack initiated from a "hard alpha" particle that formed a cavity 0.38 mm in diameter on the first cycle. A fracture mechanics analysis using FCP data showed that propagation started from the first duty cycle and the lifespan, in terms of cycles, was consistent with the 17-year age of the disk. The failure of the disk ultimately led to 111 fatalities when the plane crashed at Sioux City, Iowa, US.

While this example involved a Ti fan disk, failures of compressor disks in military aircraft having similar results have been investigated by one of the authors of this chapter (SDA). In the next section, factors affecting FCP will be considered for model systems, Waspaloy, In 718 and powder metallurgy alloys. Waspaloy and In 718 are legacy alloys still widely used for disks in jet engines. It will be demonstrated that the primary factors controlling FCP are the slip mode and environmental interactions.

1.4.5.2. *FCP in model Ni-base alloys*

1.4.5.2.1. Effects of slip mode

Previous modeling work in which FCP was viewed as a micro LCF process ahead of the crack tip resulted in the following expression [ANT 75]:

$$\frac{da}{dN} = 4\left(\frac{0.7\alpha}{E\sigma_{ys}\varepsilon_f}\right)^{\frac{1}{\beta}} \cdot \frac{1}{\ell^{\left(\frac{1}{\beta}-1\right)}} \cdot \Delta K^{\frac{2}{\beta}}$$ [1.25]

where α is a slip-mode-related constant, E is Young's modulus, σ_{ys} is the yield strength (cyclic), ε_f is the ductility (cyclic), ℓ is a process zone size in which damage accumulates ahead of a crack, and β is the Coffin-Manson exponent.

This model for FCP was verified for model materials [SAX 75] and for a broad range of engineering alloys [BAI 83] including alloys used in jet engines. The implication of the model is that planar, reversible slip will give rise to a larger process zone size, a slower rate of damage accumulation and a lower crack growth rate. It has been pointed out [MAN 06] that as good as the agreement of equation [1.25] is with LCF properties, it does not appear to be able to handle the notion that there is one striation per cycle. However, it should also be noted that other studies on striation spacing have demonstrated that such a direct relationship is not always found [HER 96].

Since slip mode and environment are clearly important factors in the FCP behavior of superalloys, a study was undertaken to sort out the effects of slip mode in Ni-base superalloys [BOW 88a, BOW 88b]. In this study, model Ni-base superalloys were used and the composition and heat treatments were manipulated to control those factors that affect the slip mode, such as anti-phase boundary energy, mismatch parameter, precipitate size and grain size. The compositions of the materials are given in Table 1.6.

Alloy Designation	Ni (w/o)	Al (w/o)	Ti (w/o)	Mo (w/o)	Cr (w/o)	B (w/o)
I	Bal	2.35	<0.01	<0.01	13.83	0.0037
II	Bal	4.92	<0.01	<0.01	14.18	0.0042
III	Bal	2.96	<0.01	<0.01	9.39	0.0037
IV	Bal	1.24	3.71	9.91	13.21	0.0060

Table 1.6. *Compositions of model Ni-base alloys*

The microstructural and compositional effects are discussed in great detail elsewhere [BOW 85]. The grain size was nominally 40 ±15 μm and the precipitate size was nominally 0.08 μm (denoted S) or 0.6 μm (denoted L). FCP testing was done at 25°C at 10 Hz over approximately three decades of crack growth. In addition to the FCP testing, LCF tests were also carried out to provide basic input and to facilitate studies of the deformation mechanisms. Microstructural parameters are provided in Table 1.7 and pertinent test results are shown in Figure 1.62.

Alloy	Grain size (μm)	γ′ size (μm)	Γ (mJ/m²)	δ (%)	V_f (%)	σ_ys (MPa)	ε_f (%)
1S	52	0.08	56	0.09	21	212	54.9
2S	51	0.09	124	0.07	27	620	36.6
3S	36	0.07	96	0.21	21	660	32.8
4S	23	0.07	420	0.18	25	660	48.4

Table 1.7. *Microstructural parameters of model alloys [BOW 88a, BOW 88b]*

The most important conclusions that could be drawn from this study were that:

– slip mode has a pronounced effect on the FCP rates at low and high values of ΔK;

– the lowest rates in Figure 1.62 were associated with precipitate shearing and planar, reversible slip (large process zone and low average crack tip strain);

– the microstructural/slip mode effects persisted even at R-ratios up to 0.7;

– crack tip closure could not explain these results. Careful experiments demonstrated that closure levels were negligible, even though the fracture surfaces associated with the lowest FCP rates were very rough and faceted. Problems with the closure concept have been demonstrated elsewhere [BOW 88c, DIN 04, VAS 04].

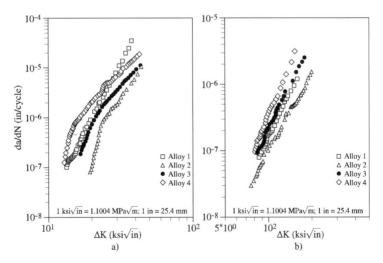

Figure 1.62. *FCP rates for model Ni-base superalloys at (a) R=0.1 and (b) R=0.7. Tested in air at 25°C and at 10 Hz [BOW 88a, BOW 88b]*

It is thus seen that the FCP rates of Ni-base alloys are strongly influenced by microstructure and the related slip mode. These alloys are susceptible to microstructural manipulation and control and those factors promoting planar, reversible slip lead to low FCP rates. Application of these principles to commercial systems are discussed in subsequent sections (e.g. 1.4.5.3.3, 1.4.5.3.5 and 1.4.5.3.6)

1.4.5.2.2. Deconvolution of environmental and slip mode effects

In the preceding section, the effects of slip mode were established. It is expected that as the temperature increases, the environment will become more important. Interactions between the environment and slip mode influence the over-all FCP behavior, especially at relatively high cycle frequencies where creep should not be a major factor. A study addressing these issues was reported [DEB 06] using the same model alloys discussed above. Testing was done at 427°C (10 Hz, $R = 0.1$) in air (A), in vacuum at 10^{-7} torr (V) and by testing in air followed by vacuum followed again by air (A/V/A). The results for alloy II-S, in which the mean boundary distance (twins or grain boundaries) was 133 µm and the precipitates were 0.024 µm diameter, are shown in Figure 1.63.

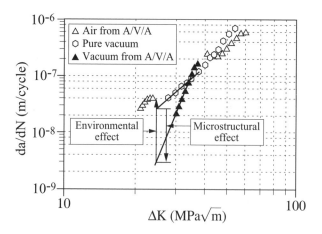

Figure 1.63. *Crack growth rates for alloy II-S in pure vacuum and for various combinations of vacuum and air. All testing was done at 427°C. Note the competing environmental and microstructural effects [DEB 06]*

The results shown in Figure 1.63 are quite instructive.

At first glance, it would appear that there is only a relatively small environmental effect by comparing the air and vacuum test results. However, another, deeper interpretation is made by comparing the A/V/A test results.

First, consider the test results from the air part of the A/V/A test compared to the vacuum portion. While it was necessary to increase the load for cracking to continue in vacuum, the test results can be extrapolated back to the point at which the change to vacuum was made. Clearly the fracture surface morphology and microstructure would be exactly identical at the changeover point, since there would not have been time for the equilibrated structure to develop. The dramatic changes in the fracture surface topology are shown in Figure 1.64 for air and vacuum.

Differences could thus be attributed to the effect of the environment, which is seen to be substantial. For these test conditions, the effect of the environment would be to increase the FCP rate from 3×10^{-9}-m/cycle in a vacuum to 4×10^{-8}-m/cycle in air – a change of over 1,200%.

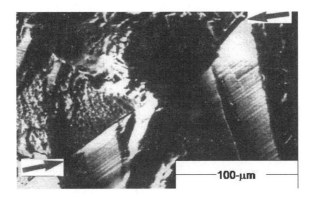

Figure 1.64. *Scanning electron micrograph of the fracture surface of II-S at 427°C. The fracture appearance in air is at the lower right and in vacuum at the upper left. Note the dramatic change in fracture surface morphology. The transition boundary is marked by the dark arrows. Transition made at 25 $MPa\sqrt{m}$ [DEB 06]*

Another important comparison can be made from these tests. If the FCP rate for the pure vacuum test is compared to that for the A/V/A test, important information about the effect of the microstructure, as manifested by the crack surface morphology, can be ascertained. In this comparison, the environment is exactly the same but the morphology of the crack surface from the A/V/A test is characteristic of that of air which is crystallographic and rough (measured roughness of 1.50). The crack surface morphology resulting from the vacuum test is, on the other hand, far less rough (measured roughness of 1.33). A crystallographic fracture surface causes a decrease from a more flat fracture surface from 2.5×10^{-8} m/cycle to 3×10^{-9} m/cycle – a reduction of 2.2×10^{-8} m/cycle (or 88%), which does not quite offset the increase due to the environment. From Figure 1.63, the net decrease on going from air to a vacuum environment is 1.5×10^{-8} m/cycle, completely consistent with the preceding discussion and calculations. Thus, the apparent small difference between air testing and vacuum testing is the net result of two competing effects: an increase due to the environment; and a decrease due to a beneficial microstructural effect. Of course, this may be viewed as two separate environmental effects: one that is intrinsic and another in which the environment changes the morphology of the fracture surface.

This insight is exceedingly important because for different materials, the balance might be struck in a very different way and it is important – both from a lifespan prediction perspective and an alloy development perspective – to understand which processes are dominant. For the alloy system, II-S, clearly both the environment and microstructural effects (environmentally induced) are important to consider.

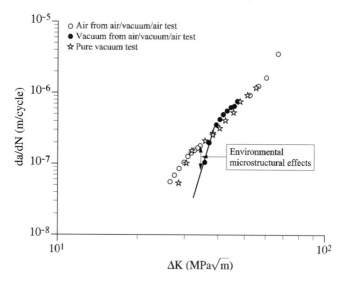

Figure 1.65. *Crack growth rates for alloy IV-S in pure vacuum and for various combinations of vacuum and air. All testing was done at 427°C [DEB 06]*

Similar tests and analyses were carried out on alloy IV-S. In this case, the fracture surface roughness was 1.59 in air at 700 K (427°C) and 1.31 in vacuum at the same temperature. Thus a larger microstructural effect would be expected because of the somewhat greater differences in the fracture surface roughness. It would be expected that the two competing effects would counterbalance. This was the case seen in Figure 1.65. In fact, for this alloy the two effects appear to almost completely counter balance each other. Consistent with this behavior, scanning electron microscopy observation of the fracture surface showed a somewhat larger difference in the morphology than was seen for II-S.

To summarize, these studies on model alloys have shown the following:

– that slip mode and fracture surface morphology are closely related;

– those factors that favor planar reversible slip (e.g. low anti-phase boundary energy, low SFE (low-stacking fault energy), low mismatch, large grains) are associated with low FCP rates,

– the environment plays two competing roles: one is intrinsic and increases the FCP rate; and the other is to increase the fracture surface roughness and reduce the FCP rate.

In the following section, the FCP of practical alloys used in jet engines will be considered and the applicability of these principles will be demonstrated.

1.4.5.3. FCP in practical Ni-base and Fe/Ni-base alloys

The previous section dealt with model alloys in which the first principles could be tested by controlling composition and heat treatment to obtain desired physical properties and microstructures. Practical Ni-base and Fe/Ni-base alloys are the result of enlightened empiricism combining intuitive understanding and vast experience.

It is important to see how the preceding principles might be applied through simple heat treatment, since changes in composition require vast development programs and certification exercises. Without changing the composition, the precipitate morphology and grain size can be controlled. At the extremes, there are basically four possible combinations of precipitate and grain size that can affect the slip mode component of the over-all crack propagation rates. These variables are discussed in the following sections for representative Ni-base alloys.

1.4.5.3.1. FCP in René 95 and Waspaloy

Some very early studies were carried out in which René 95, a powder metallurgy Ni-base superalloy used in disks, was used to demonstrate microstructural effects [BAR 77]. In these studies it was shown that improvements in FCP behavior could be brought about by changing the slip mode through varying the precipitate and grain sizes in a systematic manner. This gave perhaps the first solid indication that the principles discussed in the preceding sections could be applied to existing commercial systems.

Even more extensive studies were carried out on Waspaloy. Waspaloy is a legacy alloy that, with some minor compositional modifications, is still used as a disk material in jet engines. Waspaloy is characterized by a low lattice mismatch and by spherical precipitates. Heat treatment and deformation studies have shown that Waspaloy can be made to deform by shearing or by looping of precipitates [STO 78].

Inspired by these results, other work was carried out to vary the slip mode by producing combinations of large or small grains with large or small precipitates [LAW 85]. Thus for each grain size there were two precipitate sizes. Large grains and small precipitates favor precipitate shearing, a low average strain and slip reversibility. On the other hand, small grains and large precipitates favor precipitate looping, a high average strain and less reversible slip. Accordingly, such alloys would be expected to show extremes in FCP behavior.

Four heat treatments were carried out such that for each grain size there were two precipitate sizes (large and small) having the same strength but different slip modes. In this way, yield strength could be eliminated as a variable in the experiments. It was important to control yield strength since some false notions have been put forward that FCP depends only on modulus and yield strength. Thus for these experiments, modulus and yield strength were kept constant and the notion of FCP by macroscopic mechanical properties could be tested as well.

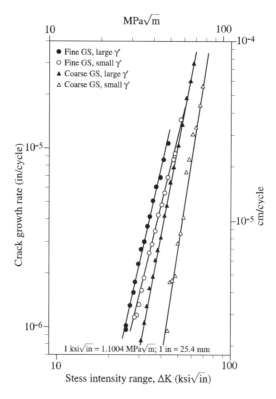

Figure 1.66. *Fatigue crack propagation of Waspaloy at constant yield strength but with different microstructures at 25°C . The FCP rates are ordered in terms of slip mode, with planar slip having the lowest rates and wavy slip having the highest rates [LAW 85]*

The results are shown in Figure 1.66. It is clear that there are very significant differences in the FCP rates. Detailed TEM work demonstrated that the alloys were ranked in one-to-one correspondence with the degree of slip planarity. The lowest rates were associated with the largest grains and smallest precipitates (deformation by precipitate shearing); whereas the highest rates were associated with the smallest grains and largest precipitates (deformation by precipitate looping). The other two were close together, as they had one factor promoting planarity and another promoting nonplanar slip, which would tend to cancel each other.

These results demonstrate an important point, which is that the FCP rates in this commercial system can be controlled by purposeful manipulation of the microstructure. The experiments were purposefully carried out at room temperature to isolate the slip mode effects that would be expected to remain relatively unaffected up to intermediate temperatures ($\sim 500°C$).

1.4.5.3.2. Fatigue threshold behavior and slip character in KM4 alloy

A detailed study of slip reversibility and its influence on the fatigue threshold was carried out on the Ni-base superalloy KM4 at 25°C, 550°C and 650°C [SHY 05]. The major contribution of this work was a new approach to defining slip reversibility (or, as they formulated it, slip irreversibility) in terms of basic dislocation processes, validated by detailed TEM and atomic force microscopy measurements over a broad range of temperatures. They developed fundamental relationships for irreversibility, denoted ϕ, and were able to show that this more fundamental physics-based definition incorporates all of the previous mesoscopic definitions of reversibility, denoted f, as unduly restrictive cases. It was recognized that irreversibility was not uniquely defined by the dislocations that exit at the surface near the crack tip as slip bands, but that dislocations are impeded in the forward and reverse directions without exiting the material. The number of dislocations exiting a surface per unit of strain, $N_{ed/\varepsilon}$, was deduced using a kind of modified Orowan equation to be:

$$N_{ed/\varepsilon} = \frac{\sqrt{2}\delta_{av}}{\Delta\varepsilon_p b} \qquad [1.26]$$

where δ_{av} is the average value of the surface upset measured by atomic force microscopy, b is the dominant Burgers vector and $\Delta\varepsilon_p$ is the plastic strain range.

Factor $\sqrt{2}$ is a geometric correction to account for the assumption that the slip band is at 45° to the loading direction. They also showed that the number of dislocations remaining in the bulk is given by:

$$N_{bd/\varepsilon} = \frac{\Delta N}{\eta}$$

[1.27]

where ΔN is the dislocation deficit (i.e. the difference between the number that would be expected and the number of dislocations that is actually emitted) and η is the fractional area swept out by dislocations on the glide plane. Using these concepts, an irreversibility parameter for the forward part of the cycle ϕ was defined as the fraction of the dislocations that have exited the crystal. In equation form, this is:

$$\varphi = \frac{\Delta N_{ed/\varepsilon}}{\Delta N_{ed/\varepsilon} + \Delta N_{bd/\varepsilon}}$$

[1.28]

Assuming that both the forward and reverse reversibility are the same, and with some additional mathematical manipulation, the irreversibility f from previous definitions is:

$$f = \phi^2$$

[1.29]

In this formulation, the degree of slip irreversibility may be measured directly from atomic force microscopy experiments. This formulation presupposes that new slip bands are not being nucleated. Measurements of ϕ were carried out on specimens that were fatigue tested at $R = 0.7$, where there could be no question of crack tip closure, whatever the validity of this concept. The results are shown in Figure 1.67.

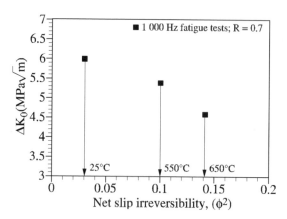

Figure 1.67. *Threshold stress intensity parameter as a function of the slip reversibility parameter for M4K superalloy. The corresponding temperature is indicated at each value of the slip irreversibility parameter [SHY 05]*

It may be more natural to think in terms of a reversibility parameter r, which may be defined in terms of ϕ, by the following equation:

$$r = 1 - \varphi^2 \tag{1.30}$$

In this formulation, Figure 1.67 shows that the threshold increases with increasing slip reversibility, as has been indicated previously based on physical understanding. This work is in substantial agreement with the FCP studies on model alloys and on Waspaloy. This is an important result since it provides further confirmation that intrinsic structural factors play an important, if not dominant, role in determining the FCP behavior at low and intermediate temperatures which are encountered for jet engine disks, notwithstanding numerous, frequently unproved, claims that so-called crack closure controls behavior.

Further confirmation of this is given in the next section.

1.4.5.3.3. FCP in the Fe/Ni-base superalloy In 718

Another important alloy used for jet engine disks is In 718. As discussed previously, this is a Fe/Ni-base alloy strengthened by coherent γ'' precipitates with a crystallographic structure different to Ni-base alloys (see Figures 1.36 and 1.37). Considerable improvements in efficiency and properties have been made with In 718 through the direct ageing process heat treatment.

A study was carried out in which a variation of the commercial heat treating process was used to obtain a 2 x 2 matrix of grain size and γ'' size at a constant strength level [KRU 87]. Testing was carried out at 427°C, where environmental factors could play a significant role.

The major practical findings of this study are shown in Figures 1.68 and 1.69. Here it is again observed that at constant strength the differences in the fatigue crack propagation rates that were seen could only be explained on the basis of microstructural differences.

For the In 718, the results were again sorted out in terms of slip planarity and reversibility with the variation in grain size having the largest effect. These intrinsic effects persisted even at an R-ratio up to 0.75 as shown in Figure 1.69.

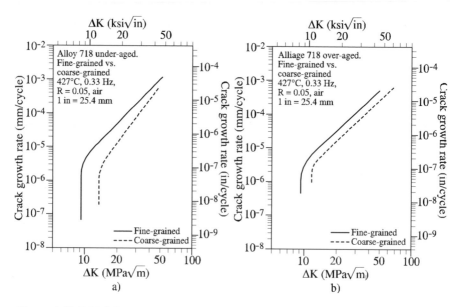

Figure 1.68. *FCP behavior of In 718 at 427°C tested at R = 0.05 and a frequency of 0.33 Hz showing the effects of two different grain and precipitate sizes at constant strength [KRU 87]*

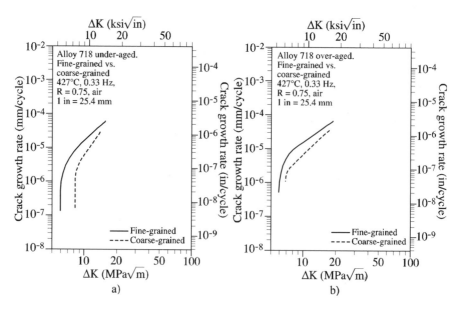

Figure 1.69. *FCP behavior of In 718 at 427°C tested at R = 0.75 and a frequency of 0.33 Hz showing the effects of two different grain and precipitate sizes at constant strength [KRU 87]*

1.4.5.3.4. Environmental, microstructural and compositional effects on the FCP rates of In 718

It has long been recognized that oxygen can have a very strong effect on the FCP rate of In 718, as shown in Figure 1.70.

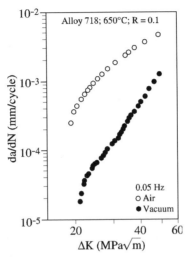

Figure 1.70. *Effect of environment on the FCP rate of In 718 at 650°C. Testing was carried out at a R-ratio of 0.1 and a cyclic frequency of 0.05 Hz under a vacuum of 5 x 10^{-8} torr [GHO 92]*

The environmental effect may manifest itself in several ways. In one particularly revealing study, the effects of oxygen partial pressure were examined [AND 92]. The operating hypothesis is that oxygen can have a short-range effect (formation of grain boundary oxides) and a long-range effect (diffusion of oxygen along boundaries with the formation of other phases that can release known embrittling agents to the grain boundaries). The basic hypothesis was that spinels of NiO and FeO can form at the grain boundary under conditions of atmospheric pressure of oxygen. However, these are fairly porous oxides and oxygen can diffuse to the interface to continue the growth of this oxide, albeit at decreasing partial pressure because of the tortuous path that oxygen must follow. With continued growth the partial pressure is no longer sufficient for the formation to be thermodynamically stable at the interface. This can be understood by straightforward appeal to a Richardson [DEH 93] diagram for oxide formation. Simply put, the partial pressure of oxygen drops below what is required for the spinel-like reactions to continue. However, at the reduced pressure Cr_2O_3 is thermodynamically stable and forms beneath the mixed oxide layer, thus passivating the surface. It was found that there was an incubation time for such passivation to occur, which increased with rising oxygen partial pressure, as seen in Figure 1.71.

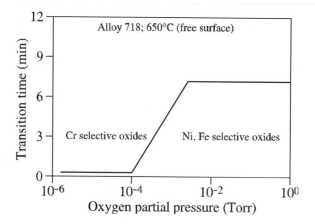

Figure 1.71. *Transition time as a function of oxygen partial pressure showing the type of stable oxide [AND 92]*

These ideas were tested by carrying out experiments for $R = 0.1$ at 650°C for three different conditions: series (a) continuous cycling; series (b) the same as continuous cycling except with a 600-second hold at minimum load; and series (c), which was the same as (b) except a small 1 Hz cycle was imposed at minimum load.

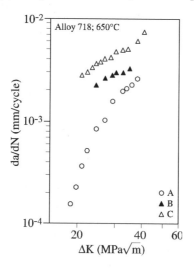

Figure 1.72. *Fatigue crack propagation rate in In 718 at 650°C for: (a) continuous cycling; (b) 600-second hold time at minimum load; and (c) same as (b) except with a 1 Hz cycle imposed at minimum load [MOL 97]*

The motivating idea was that 600 seconds was the transition time for passivation and the spinels should have stopped forming at this point. In (c), applying a small load (below the threshold so as to avoid crack advance due to "mechanical" effects) would break up the spinels at the crack tip. This would, in effect, speed up crack growth by preventing passivation and allowing the rapid spinel formation to continue at what would be effectively higher partial pressures. As can been seen in Figure 1.72, this appears to be exactly what happened and these results tend to confirm the two-step crack tip oxidation model for environmental interaction.

Additional work along these lines [MOL 97] was carried out to more precisely study the oxygen effect. In this study, it was shown that there was a minimum value (i.e. threshold) for the partial pressure of oxygen required for an environmental effect to be seen in In 718. This threshold was independent of the mechanical state, see Figure 1.73, and for a given mechanical state (i.e. ΔK) independent of the form of the cycle, see Figure 1.74.

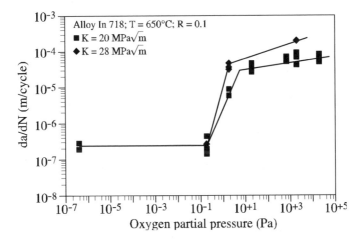

Figure 1.73. *Effect of oxygen partial pressure on the FCP rate at different ΔK levels. Note that the transition is independent of mechanical state and that after the transition the FCP rate at the plateau depends on ΔK [MOL 97]*

These figures also show that after the transition, steady state or plateau growth rates increase with increasing values and decreasing frequencies. Additional studies were carried out in which a pressure pulse of oxygen (60 seconds duration and 100 MPa pressure) was applied at various times during a 10-300-10 vacuum fatigue test. The 60-second pressure oxygen pulse allows the more damaging spinels to form without allowing time for passivation by formation of the denser Cr_2O_3.

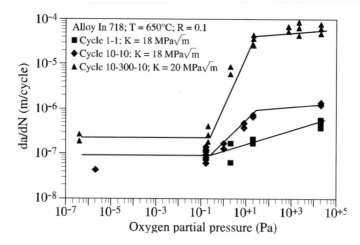

Figure 1.74. *FCP rates at a constant mechanical state but for different loading profiles. Again, the transition is independent of the loading profile but the plateau depends upon the cycle character [MOL 97]*

It is clearly shown that the crack growth rate is accelerated, regardless of where the oxygen partial pressure pulse was applied in the loading cycle, so long as the local crack tip strain rate, $\dot{\varepsilon}$, remained positive. This led to the conclusion that the damaging process of oxygen is continuous. As $\dot{\varepsilon}$ decreased (e.g. late in the hold period), the damaging effect was progressively reduced.

Finally, when the crack tip experienced a negative strain rate (start of unloading) the effect of oxygen was virtually eliminated. This led to the conclusion that any heat-treatment that would show more rapid relaxation at the crack tip (i.e. more rapid primary creep) would reduce the sensitivity of In 718 to environmental attack. This, of course, is exactly what is done for the direct ageing treatment that has coarser precipitates and a somewhat increased creep rate.

Some work was also carried out in Ni-Cr alloys with varying amounts of Cr, since increased Cr should promote the rapid formation of the more protective Cr_2O_3. The results of these studies are shown in Figure 1.75.

These results demonstrate that improvements can be made with Cr additions. However, the composition of In 718 is such that there is only limited potential for further increases in the Cr content. For a recent review on oxygen-enhanced crack growth in Ni-base alloys, see [WEI 09, WOO 06].

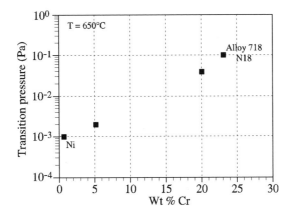

Figure 1.75. *Variation of the transition pressure with Cr for a series of Ni-Cr alloys. Alloy In 718 is included on this diagram for comparative purposes [MOL 97]*

The practical implication of these studies on In 718, taken as a whole, is that heat treatments can be devised that reduce the environmental effects and compositional modifications can be made that promote the formation of Cr_2O_3, whether directly or indirectly. Obviously, compositional changes that will directly inhibit oxidation will also improve FCP resistance to environmental effects. This is discussed at greater length below.

1.4.5.3.5.   Application   of   mechanical / environmental   model   to   Ni-base superalloy N18

These ideas were put to a practical test for developing an understanding of the FCP behavior of alloy N18, a Ni-base powder metallurgy superalloy that is used in the disks of military jet engines, such as the M88 engine in the French Rafale [CHA 97].

Samples were taken from different sections of a disk, shown in cross-section in Figure 1.76. The material and heat treatment were such that near the surface there was very fine "cooling" γ', which conferred relatively high creep resistance, whereas in the interior the γ' structure did not exhibit this variant and had less creep resistance. The cooling rates of the various regions are denoted by the letter "V" in Figure 1.76.

To examine the relationship between microstructure, deformation mode and environmental sensitivity, hold time tests were carried out in which the material was loaded to some maximum stress intensity lever, denoted Kpic, and then reduced by various percentages. The crack propagation rates in oxygen were studied as a function of the position in the disk from which the specimens were taken and the fractional load reduction. Representative results are shown in Figure 1.77.

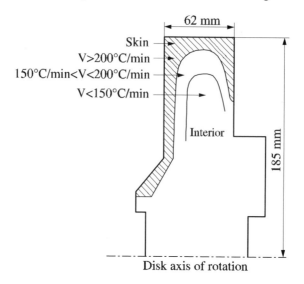

Figure 1.76. *Disk of N18 showing cooling rates at various locations. FCP samples were taken from the skin and interior regions [CHA 97]*

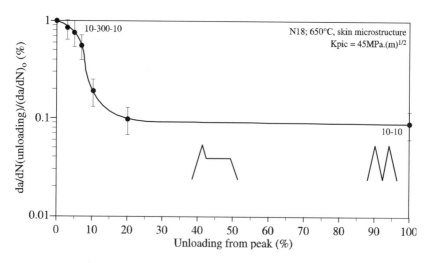

Figure 1.77. *Effect of unloading on crack propagation rate for a 10-300-10 cycle in alloy N18 at 650°C. All samples from the skin region [CHA 97]*

The results shown in Figure 1.77 are fully consistent with the notion that there is an interaction between the deformation mode and the environment through reduced stress at the crack tip. Note that as the initial load is reduced, the relative crack propagation rate decreases by about 20%, there is a plateau. That would seem to imply that further crack tip relaxation is not occurring and cracking is taking place at a steady rate. The relaxation rates of samples from the skin and interior were studied and some results are shown in Figure 1.78. In this figure, it is clear that the relaxation rate is significantly greater for the samples taken from the interior. This of course is a manifestation of a microstructure/deformation mode coupling. The finer structure in the skin area continues to relax indicating a higher creep rate and less creep resistance, at least in the initial stages.

The stress relaxation, so clearly shown in Figure 1.78, is manifested in hold time crack growth rates as shown in Figure 1.79. Numerous points can be made on the basis of this figure:

– both the interior and skin regions have identical FCP rates in vacuum that are independent of hold time;

– both regions show identical FCP rates for continuous cycling in air (10-10);

– samples from both regions show an accelerated 300-second hold time FCP rate due to the environment;

– the acceleration of the 300-second hold time FCP rate is far greater for samples taken from the skin region.

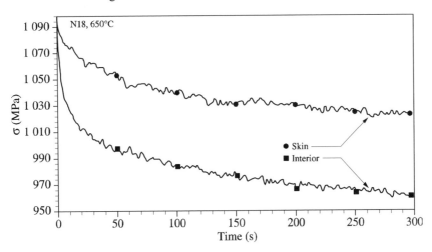

Figure 1.78. *Stress relaxation of specimens taken from the surface and from the interior of N18 at an imposed strain of 0.29% [CHA 97]*

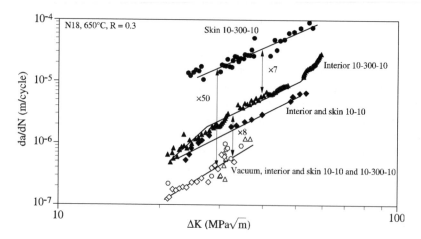

Figure 1.79. *FCP rates in N18 for various locations and conditions [CHA 97]*

These results clearly show the coupling of the environment, the microstructure and the deformation mode. They also show the fact that sensitivity to the environment may depend strongly on the microstructure through the deformation mode.

The observation of a microstructure/deformation mode/environmental effect demonstrated here is in perfect agreement with results cited previously and found elsewhere [ANT 81a, ANT 81b, ANT 80, ANT 82, ANT 83, ANT 84, ANT 92, ANT 06, LER 84, MIL 87].

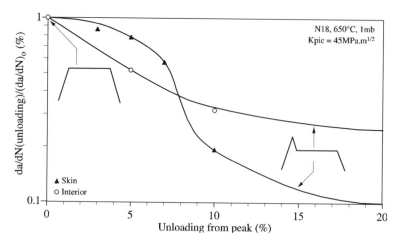

Figure 1.80. *Relative decrease in the FCP rate as a function of the load reduction for constant ΔK for samples taken from the skin and the interior [CHA 97]*

The skin and interior are compared in a somewhat different way in Figure 1.80. Here the relative FCP rates for both materials are compared for various degrees of unloading. Since the material in the interior relaxes more rapidly than the material at the skin, the effect of an external load reduction is less pronounced than it is for the material at the skin. In other words, since the material at the skin does not relax to the same extent as that at the center, an applied load drop will have a greater effect on the material at the skin than that at the interior, which has greater capacity for self-unloading (relaxation).

1.4.5.3.6. Influence of chemical composition

However appealing it may be to carry out heat-treatments to alleviate environmental degradation, some caution is in order. It must be borne in mind that during service, there is a type of continuous heat treatment being carried out that may outweigh any influence of the initial heat treatment and microstructure. For that reason, it is of interest to see whether minor chemical changes can be made that have a very large effect on the FCP behavior at elevated temperatures.

Some studies have been carried out in which direct compositional modifications were made. Since the environmental effect in In 718 is often related to grain boundary oxidation, as discussed above, chemical modifications that would suppress or eliminate grain boundary oxidation without degrading other properties would be of interest. Some studies were carried out to examine the effects of boron on FCP behavior. Boron, being an odd-sized atom, would tend to preferentially segregate at the grain boundaries. Results are shown in Figure 1.81 for testing at an R-ratio of 0.05 and a test frequency of 30 Hz.

In these studies it was shown that boron had the effect of increasing the cohesive strength of the grain boundaries, thereby shifting the "weakest link" in the crack propagation event from the oxygen-embrittled grain boundaries to fracture via a crystallographic mode involving planar, crystallographic slip. This, as has been discussed, leads to a lower crack growth rate. It was clearly shown via scanning electron microscopy examination of the fracture surfaces that the fracture mode became increasingly transgranular as the boron content increased. The effect of the boron could be to slow down diffusion at the grain boundaries. If this was the case, then the complex oxide reactions at the grain boundaries should also be slowed down.

Other studies have shown that the addition of trace amounts of P (Phosphorus) may have a beneficial effect on strengthening the grain boundaries of In 718 [SUN 98]. It was found that in the range of 0.0008–0.013% P had a beneficial effect on stress rupture and also the effect of changing the morphology of the δ phase from disks to a film-like morphlogy, with numerous faults. The beneficial effect of P was attributed to its impeding grain boundary oxidation. However, as the P concentration was increased to the 0.013–0.049% range, the creep life was lowered. This is probably because of some intrinsic effect of P in reducing the grain boundary cohesive strength. It would be logical to assume that, like B (Boron), P could be a useful addition to In 718 at trace levels in improving the creep-fatigue life.

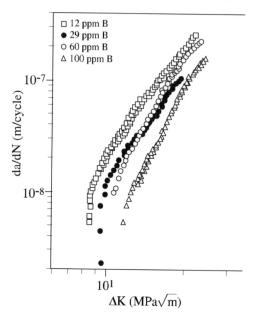

Figure 1.81. *Effect of boron on FCP rates in In 718 at 650°C. Testing was done at 30 Hz and at an R-ratio of 0.05 [XIA 06]*

It has also been shown that other modifications in the grain boundary chemistry can be made to improve the strength. In a summary report of the effects of various alloying elements on chemically modifying alloy 718 [XIE 07], it was shown that the addition of trace amounts of Mg (4–100 ppm) also are useful in increasing cyclic stress rupture lifespan probably by increasing the grain boundary strength. Again, such increases in the grain boundary strength are very likely to be reflected in improved hold-time FCP properties. The grain boundary strength is probably increased by blocking the ingress of oxygen along boundaries or possibly in tying up sulfur, as is familiar in steels. It is also mentioned that Mg, which is known to move to grain boundaries, has the effect of changing the δ grain boundary morphology from disk-like to globular. Since P additions change it to a film-like morphology, it is not at all clear that the effects of these trace elements on changing the δ morphology have any real direct effect on mechanical behavior.

1.4.5.3.7. FCP in Ni-base single crystals

Crack propagation has taken on added importance in the management and lifespan prediction of jet engine turbine blades, which generally operate at higher temperatures than disks. These, as mentioned previously, are frequently fabricated from Ni-base superalloy single crystals. While such blades have greatly extended the lifespan of turbine blades, they are very expensive. There is therefore great interest in removing them for a reason (i.e. retirement for cause, such as damage) rather than simply removing them based on a so-called safe-lifespan calculation. Thus, there is a three-fold challenge:

– to extend FCP concepts to higher temperatures;

– to apply fracture mechanics concepts to elastically anisotropic materials;

– to understand the basic mechanisms of crack propagation in these anisotropic materials.

These issues will now be addressed. In the following sections, a nominal fracture mechanics representation is used. However, a word of caution is in order about using LEFM (Linear Elastic Fracture Mechanics) to represent FCP behavior in Ni-base superalloy single crystals. Although a defect may initially be normal to the loading axis, as soon as a true crack is formed the mechanism of propagation may produce an oblique surface that calls into question the use of simple expressions for the stress intensity parameter. That is to say, the mechanism of cracking has an effect on the forcing function in such a way as to render its elementary application inappropriate. Clearly, work needs to be done to represent crack growth in terms of a more local approach, adopting a modification of what has been mentioned extensively in this chapter in the discussion of ferrous materials.

FCP in commercial single crystal alloys

FCP studies were carried out on CMSX-2, a typical superalloy used in blade applications, which have been partially reported elsewhere [ANT 92, ANT 06]. Loading was in the [001] direction. Duplicate specimens were tested as a function of temperature (25°C and 700°C), environments (air and 10^{-8} torr) and crystallographic orientation ([010] or [110] crack propagation directions with a common projected (100) crack plane). The orientations are shown in Figure 1.82 and the FCP results are reported in Figure 1.83.

It was found that the *da/dN versus* ΔK curves could not be fully understood in terms of a conventional fracture mechanics analysis representation. In particular, the changes in the fracture surface morphology were particularly challenging. Significant, long-range deviations of the crack from the plane normal to the loading axis introduced significant geometric complications into the analysis.

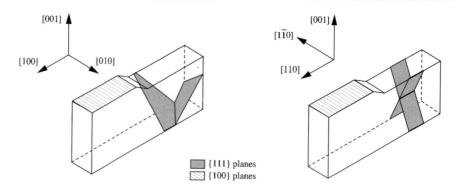

Figure 1.82. *CT (Compact Tension) specimen orientations used in FCP study of CMSX-2 [ANT 06]*

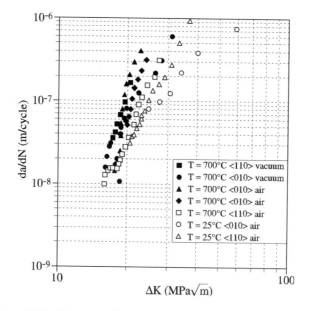

Figure 1.83. *Fatigue crack growth rates as a function of temperature, orientation and environment [ANT 06]*

The microscopic appearance of a representative fracture surface is shown in Figure 1.84 and a schematic of the fracture surface is shown in Figure 1.85. The smooth facets shown schematically in Figure 1.85 were all on {111}-type faces and as such made significant angles with the loading axis.

A detailed FEA (Finite Element Analysis) analysis including the elastic anisotropy was carried out to understand the mechanical environment in which the different cracking modes were seen. The calculation was taken from first principles and made no assumptions about the applicability of LEFM. This information was used as part of the input needed to understand these apparently baffling surface topologies. It was found that the morphology of the crack surface was either "shearing" or "precipitate avoidance", depending upon the ratio of normal stress to shear stress at the point of evaluation for a given temperature and environment.

Two distinct cracking modes were observed, and are shown in Figure 1.84.

Figure 1.84. *Typical FCP fracture surface morphologies of CMSX-2: a) precipitate shearing; and b) precipitate avoidance. The morphology was dependant upon temperature, orientation and stress ratio [ANT 06]*

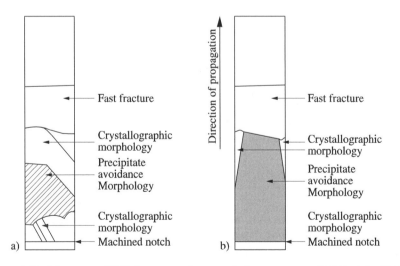

Figure 1.85. *Schematic of FCP fracture surface morphologies in air at 700°C: a) for [100]; and b) [110] crack propagation directions [ANT 06]*

Basically, precipitate shearing and large crystallographic facets that were not normal to the plane of loading were favored by low temperature, low normal-to-shear-stress ratio and vacuum. Cracking more normal to the loading axis and microscopically corrugated was observed for high values of this ratio. The complete topography of the fracture surface morphology correlated perfectly at all temperatures and for all environments by viewing transitions in terms of this stress ratio.

This surface morphology can be understood in terms of fundamental physics. For relatively low values of this ratio, the shear stress is high, favoring deformation by dislocations shearing the precipitates and ultimate crystallographic fracture. When this ratio is low, the hydrostatic component of the stress tensor is high, favoring oxygen diffusion into the material around the crack tip. This embrittles the region and leads to the "precipitate avoidance" mechanism. Such diffusion takes place more easily around boundaries, such as those between the precipitate and the matrix. Since this alloy has a high mismatch parameter, δ, oxygen ingress is favored at this location. With increasing temperature, the γ' precipitate becomes harder, as discussed previously. At a given applied shear stress, a temperature is reached where the alloy it is hard enough to resist shearing before the effects of environment and hydrostatic stress cause oxygen embrittlement. In a vacuum, the environment is far less aggressive and the fraction of crystallographic surface is much larger than in air.

Other studies have provided additional insights into cracking in single crystals. For example, FCP was examined in Mar-M200 crystals oriented with the load axis along [001], as described previously [CRO 84], at 25°C, 600°C and 850°C with crack propagation in the [100], [210] and [110] directions. It was observed that cracks would form on {111} planes at 25°C, whereas at the higher temperatures cracking was observed to be primarily on the {100} planes. The reason for this was attributed to the fact that γ' precipitates are cube-shaped and parallel to {100}. In addition, the γ' becomes harder with increasing temperature up to about 800–900°C depending on the system [EZZ 82, LAL 79, TAK 73]. Consequently, following the preceding discussion dislocations were confined to the interprecipitate channels at higher temperatures, the heavily damaged regions are macroscopically parallel to {100} and, as discussed above, subject to enhanced environmental effects leading to final fracture in these areas.

In another study [CHA 87], Mar-M200 was tested under various loading and crack propagation directions. It was found that cracking occurred on {111} planes, with the preferred plane(s) being the one(s) that had the highest and/or second highest resolved shear stress in the direction of potential Burgers vectors on the slip plane. There was considerable crack branching and the fracture surfaces were macroscopically "rough", being defined by various combinations of {111} planes.

Crack branching and surface roughness were taken into account, by correlating the crack growth rate with the effective stress intensity parameter, which takes into account the fact that modes I, II and III may be present in a single crystal that is nominally loaded in mode I.

Physically, the effective stress intensity parameter relates to the elastic strain energy, driving the crack, and may be defined mathematically as:

$$\Delta K_{eff} = \left[(\Delta K_I)^2 + \frac{C_2}{C_1}(\Delta K_{II})^2 + \frac{C_3}{C_1}(\Delta K_{III})^2 \right]^{\frac{1}{2}}$$ [1.31]

where ΔK_I, ΔK_{II} and ΔK_{III} are the mode I, II, and III stress intensity parameters and C1, C2 and C3 are elastic constants related to the anisotropy of the material and loading direction. This parameter provided acceptable correlation of the data.

It was also shown that the effect of crystallographic orientation could be understood on the basis of the amount of crack deflection and roughness-induced closure associated with a given orientation. Those orientations, for which deflection and closure were minimized, showed the most rapid crack growth rates. However, it must be recalled that closure measurement is elusive and has been shown to virtually disappear in Ni-base alloys when carefully measured [BOW 88c].

In a study of the single crystal alloy, N4 [LER 90], FCP was characterized from 25–1,000°C. The crack propagation direction was [110] and the loading direction was [001]. At room temperature, crack propagation was crystallographic with cracking occurring on {111} planes (see Figures 1.86 and 1.87).

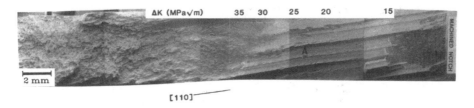

Figure 1.86. *Scanning electron microscope image of fatigue surface at 25°C. At low ΔK levels the crack propagated on and (111) planes in the [110] direction. At higher ΔK levels, the fracture surface was less macroscopically crystallographic [LER 90]*

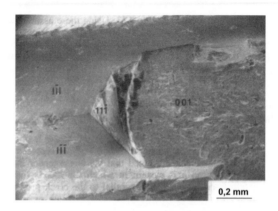

Figure 1.87. *High magnification of region A in Figure 1.86 showing crystallographic details of the fracture surface. Note the rougher, less macroscopically fractured surface in the region marked 001 [LER 90]*

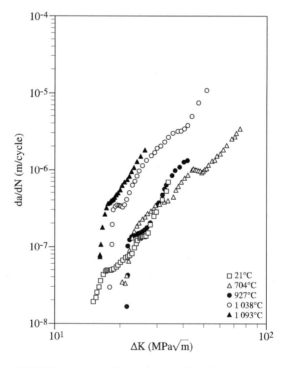

Figure 1.88. *FCP curves for N4 single crystals at different temperatures. Testing carried out in air at 0.5 Hz and an R-ratio of 0.1. Note the order of magnitude increase in FCP rate at the higher temperatures [LER 90]*

Using tensorial analysis, it was shown that the crack plane had the highest linear combination of resolved shear stress in the direction of the Burgers vector on the slip plane and normal stress across the slip plane. This criterion for multiaxial failure is motivated by the fact that the shear stresses cause dislocation motion and weaken the slip plane, while the normal stress causes the final separation. The data tended to divide themselves into two groups: slow cracking below 700°C and rapid cracking above 700°C, with fracture planes that were macroscopically normal to the loading axis and tended to follow interdendritic regions at the higher ΔK levels. At all temperatures, the crack path tended to follow the precipitate-free channels, indicative of a very high-strength precipitate. The initial increment of crack extension was on a plane inclined to the load axis. This points to the extreme care that must be exercised when interpreting results in terms of conventional fracture mechanics. The FCP results are shown in Figure 1.88.

Summary of FCP in Ni-base alloys:

– the FCP rate significantly decreases with increased grain size for both commercial and experimental alloys;

– in alloys that are strengthened by γ', decreasing the size of the γ' also decreases the FCP rate;

– at elevated temperatures the FCP rate increases significantly due to environmental interactions and slip mode effects;

– minor chemical modification can inhibit oxidation at the boundaries and significantly improve FCP rates. This has been shown to be particularly effective in In 718;

– as shown for both In 718 and N18, those microstructures in which the stresses at the crack tip relax rapidly show a correspondingly low FCP rate. This demonstrates a microstructure/deformation mode/environmental interaction;

– the FCP rate in single-crystal Ni-base alloys depends on the relative orientations of the load axis and the direction of crack. As in other alloys, the FCP rate increases with temperature;

– the FCP mechanism in single-crystal alloys is determined by a trade-off between the resistance of the γ' precipitate to shearing and environmental degradation of the γ/γ' interface;

– appropriate parameters for correlating FCP in Ni-base alloy single crystals must be more fully investigated and validated if FCP is to be analytically incorporated into the lifespan management of the high-temperature components of turbine alloys.

1.4.6. *Concluding remarks on Ni-base alloys*

It has been shown that cumulative glide behavior, damage accumulation, low cycle fatigue and fatigue crack propagation behavior are sensitive to deformation behavior, which in turn is sensitive to the details of the microstructure. At elevated temperatures, these properties are affected to a significant degree by microstructural instabilities and environmental interactions.

The alloys can be modified microstructurally, chemically and by processing to produce desired properties for applications where selected properties may be of primary importance.

Clearly a great deal of essential understanding of those factors controlling fatigue at elevated temperatures has been developed though theoretical and applied research over the past 50 years. With this sound fundamental basis, the technical community is now in a position to apply this knowledge to make more fatigue-resistant structures that can be used under more demanding conditions with a far greater degree of reliability than was thought possible only a short time ago. Two outstanding examples of this are found in the development of directionally solidified single crystals for the production of turbine blades and process control of direct aged In 718. Both of these advances depended upon our knowledge of the relationship between the microstructure, properties and processing. The field is open to even more important practical advances through exploiting the understanding that has been developed.

1.5. Lifespan prediction in high-temperature fatigue

1.5.1. *Introduction*

High-temperature fatigue is an important consideration in machinery such as jet engines, land-based turbines for power generation, pressure vessels in the power industry, nuclear reactors and others. Failure of such equipment can lead to extensive loss of life and vast economic loss, with long-term consequences for society. Thus, the mechanical integrity of these components must be assured, often for long periods of time. For example, turbine blades in jet engines can be in service for decades. Given the increasing international demands for energy and other material resources, the ability to predict and, if possible, extend their lifespan is crucial. A vast amount of literature has been published on this topic. For example, more detailed information on lifespan prediction can be found in a comprehensive review by Viswanathan [VIS 89]. In this section, only some of the most basic aspects of this topic will be discussed.

Deformation at high temperatures introduces additional damaging mechanisms that may be broadly classified as creep (i.e. time-dependent deformation) and environment (e.g. oxidation). In addition to these pure modes, complex interactions between non-time-dependent deformation (often called "plasticity" in this context), creep and environment introduce unique damage modes. Two approaches to the high-temperature fatigue problem have been developed: one based on a physical and mechanical analysis of the damage mechanisms (hereafter referred to as "physically-based models"); and another that is purely phenomenological. While the majority of the chapter has been a discussion of physically-based models, we will consider both perspectives in this section. Furthermore, we will emphasize the initiation aspect of the problem that is conveniently studied through LCF experiments.

1.5.2. *Physically-based models*

A very schematic yet attractive method for representing the various regimes of fatigue, creep and creep/fatigue interactions has been put forward by Hales [HAL 80] and is shown in Figure 1.89. In representing the cycles or time to failure as a function of the applied strain, four regions may be identified:

– in the first region, $N_{if} < N_f < N_{ic}$, where the symbols refer to the number of cycles to initiation in fatigue, the cycles to failure, and the cycles to initiate a creep crack, respectively. This simply means that fatigue failure occurs far earlier than creep damage develops and failure is essentially by transgranular fatigue, see Figure 1.89b;

– in the second region $N_{if} < N_{ic} < N_f$. In this case, there is a competition between intergranular cavitation and propagation of a fatigue crack. Basically a fatigue crack forms first, but during its propagation creep cavities begin to appear. Damage generally remains transgranular;

– in the third region, $N_{ic} < N_{if} < N_f$. Here there is a strong interaction between the creep and fatigue damage modes. The crack growth rate of the fatigue crack is (generally) accelerated by intergranular damage. Likewise, intergranular cavitation is itself accelerated by the stress and strain concentrations around the crack tip;

– in the fourth and last region, $N_{ic} < N_f < N_{if}$. Damage is primarily by creep and is completely intergranular.

Figure 1.89 is given as a schematic to help conceptualize the problem. It shows the fact that two physical approaches are appropriate, depending upon the conditions applied. The first is based on the propagation of fatigue cracks accelerated by fatigue and the second is essentially based on fatigue damage. These two methods are found throughout the technical literature. We have contributed to the development of the first kind of model. It was using this approach that made it possible to develop lifing models for 9–12% Cr steels in which oxidation plays the major role in the initiation phase with compressive hold times. It has similarly been shown that oxidation greatly accelerates crack initiation in superalloys, such as René 80 and CMSX-2. This particular mechanism is very important, since it gives rise to surface cracking as opposed to cracking in the bulk. This important distinction is not explicitly made in any of the phenomenological models that will be discussed below. High-temperature crack propagation may be accelerated by creep damage, as has been shown and modeled in austenitic stainless steels, by the environment. This is the case for all superalloys and changes in the deformation mode as have been demonstrated for all superalloys studied, notably René 95, Waspaloy and In 718.

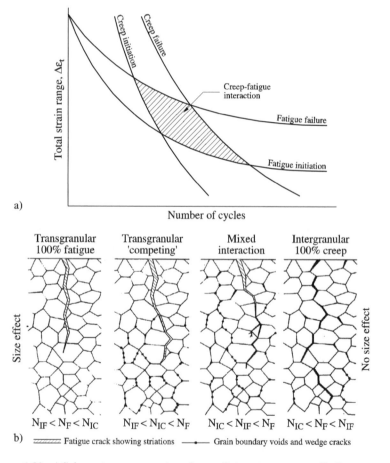

Figure 1.89. *a) Schematic representation of creep/fatigue interactions; b) illustration of physical interactions between crack and intergranular damage where it exists [HAL 80]*

Some authors, such as Skelton [SKE 93], Levaillant *et al.* [LEV 79, LEV 82] and Pineau [PIN 82, PIN 83a, PIN 83b] have shown that the propagation of a short crack can be described by an equation of the form:

$$\frac{da}{dN} = Ba^Q \qquad\qquad [1.32]$$

where Q is a constant that depends on the material and B is a function that depends on the amplitude of deformation. Recall that in Tomkins' model [TOM 68] Q is taken as 1. For creep-fatigue, the damage per cycle can be expressed as follows:

$$D = \frac{1}{N^*} = \frac{1}{N} + D_c \qquad\qquad [1.33]$$

where N^* and N are life in creep fatigue and in pure fatigue respectively.

The careful reader will note that in equation [1.33], the generic creep damage D_c must not be confused with intergranular environmental damage, written in the same manner in the section on austenitic stainless steels (equation [1.10]).

Skelton [SKE 93] has likewise shown that this linear damage accumulation law, when applied to the crack propagation rate in creep fatigue, leads to an acceleration of the crack growth rate, which may be written as:

$$\left(\frac{da}{dN}\right)^* = \left(\frac{da}{dN}\right)(1 + ND_c)^2 \qquad\qquad [1.34]$$

This is quite similar to the equation suggested by Levaillant and Pineau [LEV 82, LEV 84, LEV 88, PIN 96a, PIN 96b] (equation [1.11]), which has the following form:

$$\left(\frac{da}{dN}\right)^* = \left(\frac{da}{dN}\right)\left(\frac{1}{1 - N^* D_c}\right)^2 \qquad\qquad [1.35]$$

where D_c is now is used to denote the damage per cycle.

This model quite specifically addresses the actual damage mode in austenitic stainless steels and gives satisfactory lifespan prediction results. Other models have been developed for superalloys. We take the position that there is no universally applicable model that describes damage processes in all alloys. Instead, in developing a lifespan prediction model, each alloy must be considered in terms of the precise damage mechanisms, which will depend on the details of the material, environment and applied loading profile.

1.5.3. *Phenomenological models*

1.5.3.1. *Frequency-modified fatigue life.*

The first phenomenological model is the frequency modified fatigue life (FMFL), which was introduced by Coffin [COF 72]. In the FMFL method, cycle frequency is introduced into the basic Coffin-Manson equation by recognizing in an intuitive way that decreasing the cycle frequency will both promote creep and oxidation (not to mention their interactions) and thereby reduce the fatigue life. The original form of the equation was put in terms of the inelastic strain range in a hysteresis loop multiplied by a frequency factor that was called the frequency-modified strain range. It was later formulated in terms of the total strain, similar to work by Manson and coworkers [MAN 65], and took the following form:

$$\Delta\varepsilon_t = \Delta\varepsilon_e + \Delta\varepsilon_p = C_1\left(N_f v^{k_1-1}\right)^{-\beta_1} + C_2\left(N_f v^{k_2-1}\right)^{-\beta_2} \qquad [1.36]$$

It should be noted that in the total strain version there are six constants that may be viewed as simple fitting parameters. In an attempt to consider "unbalanced" cycles with different loading and unloading rates, a method called "frequency separation" was developed by Coffin. The tension-going and compression-going frequencies, v_t and v_c, were separated and adjustments were made for individual cases in which, for example, the more damaging nature of compressive hold time effects was recognized.

It should be recognized that with six adjustable parameters this methodology is really more of a curve fit to existing data and tends to lose contact, even with its phenomenological roots. However, it is still useful to represent experimental data in well-defined regimes. It can only be applied to lifespan prediction of components with caution, and then only in situations where there is a considerable experience base. The interested reader is referred elsewhere for a more complete discussion of this approach and its subsequent modifications [COF 72].

1.5.3.2. *Strain-range partitioning (SRP) model*

Another approach is called SRP and was introduced by the NASA group [MAN 71]. In this approach, damage is presumed to result from creep and "plasticity". Idealized cycles are shown in Figure 1.90.

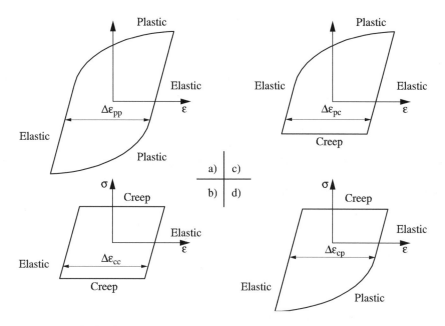

Figure 1.90. *Idealized SRP loops: a) plasticity is reversed by plasticity ($\Delta\varepsilon_{pp}$); b) creep is reversed by creep ($\Delta\varepsilon_{cp}$); c) plasticity is reversed by creep ($\Delta\varepsilon_{pc}$); and d) creep is reversed by plasticity ($\Delta\varepsilon_{cp}$). In any given cycle there may be no more than three of these components present, since it is physically impossible to have all four modes present in a cycle [MAN 71]*

The four basic types of cycles are:

– creep reversed by creep;

– plasticity reversed by plasticity;

– creep reversed by plasticity; and

– plasticity reversed by creep.

In real-world applications, the hysteresis loop is partitioned into these various modes and the lifespan is computed from the following equation:

$$\frac{1}{N_f} = \frac{F_{pp}}{N_{pp}} + \frac{F_{cc}}{N_{cc}} + \frac{F_{pc}}{N_{pc}} + \frac{F_{cp}}{N_{cp}}$$ [1.37]

In equation [1.37] the *F*s represent cycle fractions (e.g. plasticity reversed by creep) and the *N*s represent the corresponding lives taken from strain-life curves, the type of damage being considered. For example, in considering the *cp* component, the corresponding N_{pc} versus $\Delta\varepsilon_{pc}$ curve is used to determine N_{pc} as if all of the inelastic strain were of the pc type. This approach is repeated for all the other damage types present. This is an important detail that is sometimes not understood when applying this methodology and that is essential in using this method. The fractions are determined from the hysteresis loop by first accounting for the "pure" damage types pp and cc. Once these have been partitioned, the remaining component(s) are subsequently partitioned. This approach, like the FMFL, has undergone some modifications [SAL 87]. In this approach, the pc cycles are generally the most damaging in a superalloy because the compressive hold is associated with oxide cracking on the loading cycle. The pure fatigue cycle is generally the least damaging because of the reduced time for oxidation. The baseline curves are obtained experimentally with known degrees of the component of interest present. However, when the curves are obtained with uncontrolled amounts of the "other" damage form, this represents an uncontrolled variable that can lead to some dispersion in the predictions.

In another study [BAN 77], it was also demonstrated that equation [1.37] is essentially a form of Miner's law. A method has been put forward in which "pure" forms of the pc and cp curves are determined by extrapolation. Using this technique and accounting for true interactions (Miner's law constant $\neq 1$) markedly improved the accuracy of the prediction.

The same factors that were considered in developing the FMFL and SRP approaches occur at the tips of cracks, so these concepts may be carried over to the case in which the main interest is the propagation of a fatigue crack. Whereas strains are directly measured during traditional creep-fatigue tests, such is not the case for crack propagation. Thus, some kind of indirect approach is required to numerically compute the inelastic strains, both in terms of magnitude and type. This requirement obviously makes it more difficult to apply the SRP approach to FCP than to smooth bar fatigue.

1.5.3.3. Damage accumulation concepts

The basic idea of these phenomenological models is to accumulate two elementary damage modes, the first one (related to fatigue, D_f) which is independent of time, and the second one (related to creep and/or oxidation, D_c) which is time dependent. In the literature, several damage accumulation rules have been proposed that will now be discussed.

1.5.3.3.1. Linear damage rule

The simplest law describing damage accumulation is known as the time and cycle fraction rule and is written as:

$$D = D_f + D_c = \frac{n}{N_f} + \frac{t}{t_R} \qquad [1.38]$$

where n is the number of cycles under a set of conditions, N_f is the fatigue life for the imposed conditions, t is the time for which the component of interest is under load and t_R is the time for creep rupture under these conditions.

As has been discussed previously, it is not always straightforward to select either the appropriate time in a cycle or the total time to creep rupture, since the conditions vary during the course of a cycle. That having been said, there are instances in which $D = 1$ at failure in equation [1.38]. However, that might, in some sense, be regarded as fortuitous.

As seen in Figure 1.18, there are cases in which $D < 1$ at failure. For example in tests for which fatigue is followed by creep, there are cases in which D is significantly greater than 1. Equation [1.38] is the basis for the Robinson-Tiara rule [ROB 52, TAL 62]. The time and cycle fraction rule is shown as an unbroken, straight line in Figure 1.91.

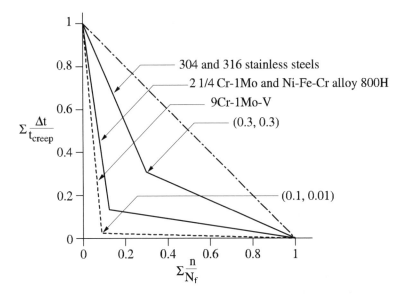

Figure 1.91. *Cumulative damage diagram of fatigue damage D_f versus macroscopic creep damage, D_c. In addition to the linear damage rule (straight line A–B) two other bi-linear damage laws are represented by two broken lines. These are the bi-linear rules recommended by the ASME code for austenitic stainless steels (304SS and 316SS) and for 2 1.4 Cr-1 Mo steels and 800H. Finally, the bi-linear rule for 9Cr-1 Mo is also shown*

Very significant negative deviations from the linear damage accumulation law are frequently seen for reasons that will be explained. Thus, in practice, bi-linear damage accumulation equations are frequently used, such as those indicated in Figure 1.91.

These deviations from the simple linear damage rule of equation [1.38] have several causes. The principal reason for such deviations is related to the fact that in order to define a reference state, N_f and t_R virgin material is tested. Thus, all microstructural changes caused by creep-fatigue are ignored. Moreover, if this rule is applied to fatigue relaxation tests, the relaxed stress is different for each cycle, depending on whether the material hardens (e.g. austenitic stainless steels) or softens (e.g. 9–12% Cr steels). For superalloys the situation can be even more complicated (hardening followed by softening).

Finally, for hold times at maximum strain the creep damage, as previously mentioned, must be defined in a state of continually decreasing stress. The simplest damage parameter for these conditions has been defined by Spera [SPE 73] and uses a simple integration during the hold period:

$$D_c = \int\limits_0^{t_h} \frac{dt}{t_R\left(\sigma_T(t)\right)} \tag{1.39}$$

where $\sigma_T(t)$ is the time-dependent stress during the hold period at some temperature T.

Some researchers prefer to compute the fatigue damage D_c in terms of strain rather than in terms of time. The damage increment is thus represented by:

$$D_c = \frac{d\varepsilon_e}{\varepsilon_{f,T}\left(\dot{\varepsilon}\right)} \tag{1.40}$$

where ε_f is the creep ductility at temperature T and a viscoplastic strain rate $\dot{\varepsilon}$.

This is so we can introduce the fact that in many materials creep ductility decreases with the creep strain rate. This is referred to as "creep ductility exhaustion". Another reason for this formulation is related to the fact that multiaxial creep damage is more easily formulated in terms of strain. For example, Spindler [SPI 04] has suggested the following:

$$\frac{\varepsilon_f^{multi}}{\varepsilon_f^{uni}} = \exp\left[p\left(1-\frac{\sigma_1}{\sigma_{eq}}\right)+q\left(\frac{1}{2}-\frac{3}{2}\frac{\sigma_m}{\sigma_{eq}}\right)\right] \tag{1.41}$$

This equation is completely analogous to that recently suggested by Auzoux [AUZ 05], who algebraically manipulated equation [1.41] to give:

$$\frac{\varepsilon_f^{multi}}{\varepsilon_f^{uni}} = \exp\left[p\left(\frac{1}{3} - \frac{\sigma_m}{\sigma_{eq}}\right)\right] \qquad [1.42]$$

As has already been stated, significant deviations from the linear damage accumulation rule are observed. These deviations prompted some researchers to suggest phenomenological, nonlinear, damage accumulation expressions.

1.5.3.3.2. Nonlinear damage accumulation

In Chapter 2 of [BAT 11], a mechanics-based nonlinear damage accumulation formulation has been developed in a detailed fashion for continuous media. Here we will simply outline the formalism developed by Kachanov [KAC 58] and by Lemaitre and Chaboche [LEM 85].

In a formal way, a process involving two damage mechanisms may be described by two total differentials:

$$dD_{fat} = f_{fat}\left(\sigma_M, \bar{\sigma}, T, D_f + D_c\right)dN \qquad [1.43]$$

$$dD_c = f_{cr}\left(\sigma, T, D_f + D_c\right)dt \qquad [1.44]$$

where for the uniaxial case σ_M is the maximum stress in a cycle, $\bar{\sigma}$ is the mean stress, T is the temperature and the other terms have been previously defined. Creep-fatigue coupling is introduced by writing that the increment for each damage mode depends on the current total damage $(D_f + D_c)$ at cycle N or time t. The expressions introduced for the functions of damage f_{fat} and f_{cr} have been given explicitly in [CHA 11].

A model in the same vein was developed by Maiya and Majundar [MAI 77]. Their model, which is essentially a model for fatigue crack growth accelerated by creep cavitation, has the following form:

$$\frac{1}{a}\frac{da}{dN} = \left\{\frac{T}{C}\right\}\left|\varepsilon_p\right|^m \left|\dot{\varepsilon}_p\right|^k \left(1 + \ln\frac{c}{c_o}\right) \qquad [1.45a]$$

$$\frac{1}{c}\frac{dc}{dt} = \left\{\frac{G}{-G}\right\}\left|\varepsilon_p\right|^{m_c} \left|\dot{\varepsilon}_p\right|^{k_c} \qquad [1.45b]$$

The notation between parentheses in equation [1.45a] indicates in tension the RHS (Right Hand Side) is to be multiplied by T and in compression by C. Similarly, for equation [1.45b]. Thus, the first term is not a true fraction but a symbolic guide.

Fatigue damage is represented by crack length, whereas creep damage is represented by c, the fraction or size of cavities. In these equations, the absolute values of strain and strain rate are used. However, other quantities such as T, C, G, k, k_c, m, m_c, a_o (initial crack length) and c_o (initial cavity size) are determined experimentally. If it is assumed that $m = m_c$ in symmetric cycles in strain control and in continuous cycling without hold time, these equations take on the following form, which can be used for integration:

$$\left(\frac{da}{dN}\right)_{CF} = \left(\alpha \frac{T+C}{m+1}\right)\left(\Delta\varepsilon_p\right)^{m+1}\left(\dot{\varepsilon}_p\right)^{k-1} \qquad [1.46]$$

This equation is analogous to that suggested by Tomkins [TOM 75] from a totally different starting point, with the difference that in equation [1.46] the strain rate appears explicitly, which is not the case for Tomkins' expression. For creep-fatigue cycles with hold times at the maximum load (strain) this equation takes on the following form:

$$\left(\frac{da}{dN}\right)_{CF} = \left(\frac{da}{dN}\right)_{PF}\left(1+\gamma G I N\right) \qquad [1.47]$$

where: $I = \int\limits_{cycle} \left|\varepsilon_p\right|^m \left|\dot{\varepsilon}\right|^{k_c} dt$

Here, N is the number of cycles and γ is a coefficient that related to parameter G. This equation, completely like the one we introduced during the development of physically-based models for interactions between fatigue and intergranular creep damage (equation [1.11]) for stainless steels, shows a multiplicative effect of creep damage on the crack growth rate.

1.6. Conclusions

During the past 50 years, high-temperature fatigue has been the subject of numerous studies. The approach that has been taken in this chapter has been to concentrate on three important classes of materials:

– 9–12% Cr steels;

– austenitic stainless steels;

– Ni-base superalloys.

This limitation admittedly gives only a partial overview of the field of high-temperature fatigue. However, these three classes of materials are among the most used, if not *the* most used, at high temperature. They have the additional advantage of illustrating a wide range of damage mechanisms and the principles reviewed and developed here can be extended to other materials with minor modifications.

The references in this chapter, at least in a qualitative sense, are quite representative of research in this field. In examining these nearly 200 references, it is striking that 50 (i.e. more than one-quarter) were published in the 1980–85 period. It is equally striking that the distribution of publication dates is not simply Gaussian centered about these dates. From about the year 2000, the number of publications per year has shown an increase to 30 from 2003–08.

No doubt there are several reasons for the form of this histogram. For example, you might argue that, among other non-technical factors, it reflects our age. However, it is our view that the peak seen in the 80s is really a reflection of industrial development in the jet engine and nuclear reactor industries, especially fast neutron reactors. The same kind of explanation can be invoked to explain the recently increased activity in high-temperature fatigue research. For example, solving the energy crisis requires more efficient "machinery". Based on fundamental thermodynamics, increased efficiency requires higher operating temperatures. It is worth recalling the simple but profound thermodynamic formula for the efficiency of a thermal process:

$$Eff(\%) = \left(1 - \frac{T_a}{T_h}\right) \cdot 100 \qquad\qquad [1.48]$$

where T_h and T_a are the operating temperature and ambient temperature, respectively. We are thus inexorably driven by this immutable law to:

a) better understand how materials behave so they can be used with confidence at their true limits; and

b) to develop even more temperature-resistant materials which, of course, are subject to the process indicated in a).

Similar ideas apply to conventional heat engines. For example, had we included aluminum alloys, we would have shown a significant increase in work being done on cast Al-Si alloys, which are widely used in various types of internal combustion engines.

In this chapter we have illustrated the following salient features of high-temperature fatigue:

– modifications in the mode of deformation by heat treatments and composition changes to favor desirable deformation modes;

– damage by oxidation;

– intergranular damage due to creep.

When the temperature is increased in the absence of environmental or creep effects, modifying the mode of deformation is generally a very effective approach. This has been shown for both model and industrial alloys. It has been clearly shown that for pre-existing cracks (i.e. crack propagation), changing the chemistry and heat treatment can lead to favorable deformation modes and longer propagation lives. However, as the temperature is increased other factors play a role, especially in the initiation of fatigue cracks. With increased temperature, there is a tendency for more homogeneous slip, which delays the onset of fatigue crack initiation. In many instances, depending on the application, most of the lifespan is spent in crack initiation, so slip homogenization is desirable. There are even certain cases in which fatigue properties improve with increasing temperature.

However, in most cases increasing the temperature reduces the lifespan of the material. Quite often, this decreased fatigue life is directly related to oxidation, which is frequently intergranular. This is particularly the case for most Ni-base superalloys. Oxidation effects can give rise to some surprising results, such as a reduction in fatigue life for compressive hold times as compared to tensile hold times. Such effects are seen in certain superalloys as well as in 9–12% Cr steels. These oxidation effects are surface-related, which means that this particular localized damage mode must be taken into account when developing models. It was shown that for Ni-base alloys, an oxidation-base model that takes into account the interaction between the deformation mode and oxide formation can be developed. The negative effects of oxide spiking on fatigue life can be offset in high mismatch parameter alloys by rapid deformation-induced coarsening, which reduces the stress in strain-control fatigue. We have shown this trade-off quite clearly for René 80, where for a given strain range the lifespan decreased with increasing temperature (stable microstructure) and then increased for higher temperatures (stress-induced coarsening).

On the other hand, intergranular creep damage is dominant in austenitic stainless steels. It was shown that fatigue life reductions during hold-time fatigue in these materials are linked to this damage mode (which is easily measured, as has also been shown). Intergranular creep, in contrast to oxidation, occurs throughout the bulk of the material. Measurements of the degree of intergranular cavitation provide a basis for developing physically-based lifespan prediction models.

Creep-environment effects are particularly important in Ni-base superalloys. In these materials, oxidation at the surface promotes crack initiation but also significantly affects crack propagation. While details of the local damage mechanisms are not yet fully understood, overall general ideas based on these physical processes are certainly useful in developing better alloys. In particular, the crack growth rate in Ni-base alloys is reduced when the microstructure is such that crack-tip relaxation can occur. The nature of the oxide is also of fundamental importance and may be controlled through compositional modification. For example, in In 718 great improvements can be brought about with increased boron and chromium content. These modifications either inhibit oxidation (boron) or provide a protective layer (chromium). However, other metallurgical considerations limit the extent to which the composition may be modified. In the case of In 718, it appears as if the chromium composition is near that limit.

The search for higher operating temperatures in jet engine turbine blades has led to several technical innovations. For example, thermal barrier coatings provide shielding that limits oxidation effects. The compatibility of the coefficients of thermal expansion between coating and base metal must be borne in mind, since mismatches can lead to new problems, especially in thermomechanical fatigue. In this area, the adhesion between coating and base metal is important and transition zones between the metal and coating are desired, as opposed to discontinuous interfaces. It has also been shown, especially at low and intermediate temperatures, that coating cracks develop and proceed directly into the matrix. When this happens, all of the principles that have been discussed for the base metal come into play.

The primary goal of this chapter was to examine physical damage processes and models. However, a more modest part was devoted to a review of lifespan prediction methods. Following the lifespan prediction approach, we thought it was important to show that the search for a single, all-encompassing model is doomed to failure. Lifespan prediction must be based on actual processes. As has been shown, such processes are very different in terms of their nature and evolution with cycles and time. Existing phenomenological models are widely used and were also discussed. Great care must be exercised when using the phenomenological models to be sure that they correspond to the physical reality of actual damage processes. Their use requires validation and they should never be extrapolated to conditions for which they have not been developed. Viewing such models in terms of basic physical processes is the only way in which their regime of validity can be established.

1.7. Acknowledgments

The authors would like extend their heart-felt thanks to Joëlle for her help in the preparation of this chapter.

1.8. Bibliography

[ALE 04a] F. ALEXANDRE, Aspects probabilistes et microstructuraux de l'amorçage des fissures de fatigue dans l'alliage INCO 718, PhD Thesis, Ecole des Mines de Paris, 2004.

[ALE 04b] F. ALEXANDRE, S. DEYBER, A. PINEAU, "Modelling of the optimum grain size on the low cycle fatigue life of a Ni based superalloy in the presence of two possible crack initiation sites", *Scripta Mater.*, vol. 50, pp. 25-30, 2004.

[AND 92] E. ANDRIEU, R. MOLINS, H. GHONEM, A. PINEAU, "Intergranular crack tip oxidation mechanisms in nickel-based superalloy", *Mat. Sci. Eng.*, vol. A154, p. 21-28, 1992.

[ANT 75] S.D. ANTOLOVICH, A. SAXENA, G.R. CHANANI, "A model for fatigue crack propagation", *Eng. Fract. Mech.*, vol. 7, pp. 649-652, 1975.

[ANT 79a] S.D. ANTOLOVICH, P. DOMAS, J.L. STRUDEL, "Low cycle fatigue of René 80 as affected by prior exposure", *Met. Trans,* vol. 10A, pp. 1859-1868, 1979.

[ANT 79b] S.D. ANTOLOVICH, S. BASHIR, P. TAUPIN, "Low cycle fatigue of as-HIP and HIP + forged René 95", *Met. Trans.*, vol. 10A, pp. 1481-1490, 1979.

[ANT 80] S.D. ANTOLOVICH, R. BAUR, S. LIU, "A mechanistically based model for high temperature low cycle fatigue of nickel base superalloys, Superalloys", in: J.K. TIEN, S.T. WLODEK, H. MORROW III, M. GELL, G.E. MAUER (eds.), *Superalloys 1980*, American Society for Metals, Champion, PA., p. 605-613, 1980.

[ANT 81a] S.D. ANTOLOVICH, S. LIU, R. BAUR, "Low cycle fatigue behavior of René 80 at elevated temperature", *Met. Trans.*, vol. 12A, pp. 473-481, 1981.

[ANT 81b] S.D. ANTOLOVICH, E. ROSA, A. PINEAU, "Low cycle fatigue of René 77 at elevated temperatures", *Mat. Sci. Eng.*, vol. 47, pp. 47-57, 1981.

[ANT 82] S.D. ANTOLOVICH, J.E. CAMPBELL, "Fracture properties of superalloys, application of fracture mechanics for the selection of metallic structure material", in: J. CAMPBELL, W.W. GERBERICH, J.H. UNDERWOOD (eds), *American Society for Metals*, Metals Park, Ohio, pp. 253-310, 1982.

[ANT 83] S.D. ANTOLOVICH, N. JAYARAMAN, "The effect of microstructure on fatigue behavior of nickel base alloys", in: J.J. BURKE, V. WEISS (eds), *Fatigue Environment and Temperature Effects*, Plenum Press, New York, pp. 119-144, 1983.

[ANT 84] S.D. ANTOLOVICH, N. JAYARAMAN, "Effects of environment and structural stability on the low cycle fatigue behavior of nickel base superalloys", *High Temperature Technology*, vol. 21, pp. 3-13, 1984.

[ANT 89] S.D. ANTOLOVICH, B. LERCH, "Cyclic deformation and fatigue in Ni-base alloys", s, in: J.K. TIEN, T. CAUFIELD (eds), *Superalloys, Supercomposites and Superceramic*, Academic Press, New York, pp. 363-411, 1989.

[ANT 92] B.F. ANTOLOVICH, A. SAXENA, S.D. ANTOLOVICH, "Fatigue crack propagation in single crystal CMSX-2 at elevated temperature", in: S.D. ANTOLOVICH, R.W. STUSRUD, R.A. MACKAY, D.L. ANTON, T. KHAN, R.D. KISSINGER (eds), *Superalloys 1992*, pp. 727-736, 1992.

[ANT 00] S.D. ANTOLOVICH, *Tension Testing Methods in Materials Research*, John Wiley and Sons, New York, 3a.1-3a.1.10, 2000.

[ANT 02] S.D. ANTOLOVOICH, A. SAXENA, "Thermomechanical fatigue: Mechanisms and practical life analysis", in: W.T. BECKER, R.J. SHIPLEY (eds), *ASM Handbook, vol. 11, Failure Analysis and Prevention*, ASM International, Materials Park, Ohio, pp. 738-745, 2002.

[ANT 06] S.D. ANTOLOVICH, B.F. ANTOLOVICH, "The effects of microstructure, deformation mode and environment on fatigue", *Proceedings of the 16th European Conference of Fracture*, Alexandroupolis, Greece, pp. 1245-1246, July 3-7, 2006.

[ARA 01] N.K. ARAKERE, E. OROZCO, "Analysis of low cycle fatigue properties of single crystal nickel-base turbine blade super-alloys", *High Temperature Materials and Processes*, vol. 20, pp. 403-419, 2001.

[ARG 96a] D. ARGENCE, Endommagements couplés de fatigue et de fluage sous chargement multiaxial appliqué à un acier inoxydable austénitique, PhD Thesis, Ecole des Mines de Paris, 1996.

[ARG 96b] D. ARGENCE, A. PINEAU, "Predictive metallurgy applied to creep-fatigue damage of austenitic stainless steels", in: E.H. HONDROS and M.MC LEAN (eds), *Structural Materials: Engineering Application Through scientific Insight*, The Institute of Materials, The University Press, Cambridge, pp. 229-257, 1996.

[ARM 04] A. ARMAS, C. PETERSEN, R. SCHMITT, M. AVALOS, I. ALVAREZ, "Cyclic instability of martensitic laths in reduced activation ferritic/martensitic steels", *J. Nucl. Materials*, Vol. 329-333, pp. 252-256, 2004.

[AUZ 05] Q. AUZOUX, L. ALLAIS, C. CAES, B, GIRARD, I. TOURNIE, A.F. GOURGUES, A. PINEAU, "Intergranular damage in AISI 316L (N) austenitic stainless steel at 600°C: Prestrain and multiaxial effects", *Nucl. Eng. & Des*, vol. 235, pp. 2227-2245, 2005.

[BAI 83] J.P. BAILON, S.D. ANTOLOVICH, "The effect of microstructure on fatigue crack propagation: a review of existing models and suggestions for further research, Fatigue Mechanisms: Advances in quantitative measurement of physical damage", in: J. LANKFORD, D.L. DAVIDSON, W.L. MORRIS, R.P. WEI (eds), *ASTM STP 811*, pp. 313-349, 1983.

[BAN 77] P.C. BANIA, The high temperature low cycle fatigue behavior of a near-alpha titanium alloy, PhD Thesis, University of Cincinnati, 1977.

[BAR 77] J. Bartos, S.D. Antolovich, "Effect of grain size and gamma prime size on fatigue crack propagation in René 95", in: D.M.R. Taplin (ed.), *Fracture 1977*, vol. 2, University of Waterloo Press, Canada, pp. 996-1006, 1977.

[BAT 10] C. Bathias, A. Pineau, *Fatigue of Materials and Structures: Fundamentals*, ISTE, London and John Wiley & Sons, New York, 2010..

[BAT 11] C. Bathias, A. Pineau, *Fatigue of Materials and Structures: Application to Design*, ISTE, London and John Wiley & Sons, New York, 2011.

[BLO 85] T. A. Bloom, U.F. Kocks, P. Nash, "Deformation behavior of 4; Mo alloys", *Acta Met.*, vol. 33, pp. 265-272, 1985.

[BOU 77] J. Bourgeot, B. Henry, B. Michaut, B. Thomas, "Amélioration de la tenue au fluage d'aciers inoxydables austénitiques stabilisés. Etude du rôle de la précipitation, du bore et du traitement d'hypertrempe", *Mémoires Scientifiques, Revue de Métallurgie*, pp. 491-512, 1977.

[BOU 81] B. Boursier, Evaluation of damage mechanisms in the nickel base superalloy Rene 80 under low cycle fatigue in the temperature range 75F-1400F, MS Thesis, University of Cincinnati, 1981.

[BOW 85] R. Bowman, Effect of melt spinning on the microstructure and mechanical properties of three nickel base superalloys, MS Thesis, Georgia Institute of Technology, 1985.

[BOW 88a] R.R. Bowman, The effect of microstructure on the fatigue crack growth resistance of nickel base superalloys, PhD Thesis, Georgia Institute of Technology, 1988.

[BOW 88b] R.R. Bowman, S.D. Antolovich, The effect of microstructure on the fatigue crack growth resistance of nickel base superalloys, in: D.N. Duhl, G. Maurer, S. Antolovich, C. Lund and S. Reichman (eds), *Superalloys 1988*, AIME, Warrendale, PA, pp. 565-574, 1988.

[BOW 88c] R.R. Bowman, S.D. Antolovich, R.C. Brown, "A demonstration of problems associated with crack closure measurement techniques", *Eng. Fract. Mech.*, vol. 31, pp. 703-712, 1988.

[BRA 91] J.C. Brachet, Alliages martensitiques 9Cr–1Mo: effets de l'addition de l'azote, du niobium et du vanadium sur la microstructure, les transformations de phase et les propriétés mécaniques, PhD Thesis, University of Paris-Sud, Orsay, 1991.

[BRE 81] J. Bressers, B. Verhegghe, "The effect of time-dependent processes on the high-temperature LCF life of Waspaloy", *Res Mechanica Letters*, vol. 1, pp. 55-59, 1981.

[BRE 83] J. Bressers, M. Roth, "Advances in life prediction methods", in: D.A. Woodford and J.R. Whitehead (eds), *ASME Int. Conf. on Advances in Life Prediction Methods*, ASME, New York, pp. 85-92, 1983.

[BRU 06] A. Bruckner-Foit, S. Huang, "Numerical simulation of micro-crack initiation of martensitic steel under fatigue-loading", *Int. J. Fatigue*, vol. 28, pp. 963-971, 2006.

[CHA 81] K.D. CHALLENGER, A.K. MILLER, C.R. BRINKMAN, "An explanation for the effects of hold periods on the elevated temperature fatigue behaviour of 2¼Cr – 1Mo steel", *J. Eng. Mater. Technol.*, vol. 103, pp. 7-14, 1981.

[CHA 87] K.S. CHAN, G.R. LEVERANT, "Elevated temperature fatigue crack growth behavior of MAR-M200 single crystals", *Met. Trans.*, vol. 18A, pp. 593-602, 1987.

[CHA 97] J.C. CHASSAIGNE, High temperature crack propagation in Ni-base superalloy N18 fabricated by pm; study of the coupling of mechanical and environmental effects at the head of a crack, PhD Study, Ecole des Mines de Paris, 1997.

[CHA 11] J.L. CHABOCHE, "Cumulative damage", in C. BATHIAS and A. PINEAU (eds), *Fatigue of Materials and Structures: Application to design*, ISTE, London, John Wiley & Sons, New York 2011.

[CHU 76] D.W. CHUNG, M. CHATURVEDI, D.J. LLOYD, "The characteristics of the barrier during serrated yielding in the γ' precipitation hardened superalloys (Co-Ni-Cr system)", *Acta Met.*, vol. 24, pp. 227-232, 1976.

[CLA 80] M.C. CLAVEL, C. LEVAILLANT, A. PINEAU, "Influence of micromechanisms of cyclic deformation at elevated temperature on fatigue behavior", in: R.M. PELLOUX and N.S. STOLOFF (eds), *Creep-Fatigue-Environment-Interactions*, AIME, Warrendale, PA, pp. 24-45, 1980.

[CLA 82] M.C. CLAVEL, A. PINEAU, "Fatigue behavior of two nickel base alloys - Part I: Experimental results on low cycle fatigue, fatigue crack propagation and substructures", *Mat. Sci. & Eng.*, vol. 55, pp. 157-171, 1982.

[COF 54] L.F. COFFIN JR, "A study of the effect of cyclic thermal stresses on a ductile metal", *Trans. ASME*, vol. 76, pp. 931-950, 1954.

[COF 71] L.F. COFFIN JR, "The effect of frequency on the cyclic strain and low cycle fatigue behavior of cast Udimet 500", *Met. Trans. B*, vol. 2, pp. 3105-3113, 1971.

[COF 72] L.F. COFFIN, "Fatigue", *Annual Review of Materials Science*, vol. 2, R.A. HUGGINS, Annual Reviews Inc., 313-348, 1972.

[COF 74] L.F. COFFIN, "The effect of frequency on the cyclic strain and fatigue behavior of cast René at 1600°F", *Met. Trans.*, vol. 5, pp. 1053-1060, 1974.

[CRO 84] J.S. CROMPTON, J.W. MARTIN, "Crack growth in a single crystal superalloy at elevated temperature", *Met. Trans.*, vol. 15A, pp. 1711-1719, 1984.

[DEB 06] A. DE BUSSAC, S.D. ANTOLOVICH, "The effects of microstructure and environment on the fatigue crack propagation behavior of a Ni-base alloy", *Proceedings of the 9th International Fatigue Congress*, Atlanta, GA, Published in CD format as Oral Reference FT 506, Elsevier, May 14-19, 2006.

[DEH 93] R.T. DEHOFF, *Thermodynamics in Materials Science*, McGraw-Hill, New York, 1993 (for an excellent discussion of metal oxide reactions and uses of the Ellingham-Richardson diagram).

[DIE 84] D. DIEDERICH, B. LERCH, S.D. ANTOLOVICH, "Observations of low cycle fatigue and fatigue crack propagation substructures of Waspaloy", *Proceedings of the VIII Inter-American Conference on Materials*, pp. 7.1-7.6, June 25-29, 1984.

[DIN 04] S. DINDA, D. KUJAWSKI, "Correlation and prediction of fatigue crack growth for different R-ratios using Kmax and ΔK+ parameters", *Eng. Fract. Mech*, vol. 71, pp. 1779-1790, 2004.

[DUH 89] D.N. DUHL, "Single crystal superalloys, superalloys, supercomposites, and superceramics", in: J. K. TIEN and T. CAUFIELD (eds), *Superalloys, Supercomposites and Superceramics*, Academic Press, New York, pp. 149-182, 1989.

[EGG 05] G. EGGELER, A. DLOUHY, "Boron segregation and creep in ultra-fine grained tempered martensitic ferritic steels", *Zeitschrift für Metallkunde*, vol. 7, pp. 743-748, 2005.

[EGG 87] G. EGGELER, N. NILSVANG, B. ILSCHNER, "Microstructural changes in a 12% chromium steel during creep", *Steel Research*, vol. 2, pp. 87-103, 1987.

[EGG 89] G. EGGELER, "The effect of long-term creep on particle coarsening in tempered martensite ferritic steels", *Acta Metall.*, vol. 37, pp. 3225-3234, 1989.

[ETI 73] C.F. ETIENNE, W. DORTLAND, H.B. ZEEDIJK, "On the capability of austenitic steel to withstand cyclic deformation during service at elevated temperatures. Creep and fatigue in elevated temperature application", paper C225/73, *Int. Conf., Creep and Fatigue in elevated temperature application*, Philadelphia, 1973.

[EZZ 82] S.S. EZZ, V. PAIDAR, "The tension/compression flow stress asymmetry of Ni3 (Al,Nb) single crystals", *Acta Met.*, vol. 30, pp. 921-926, 1982.

[FOU 77] D. FOURNIER, A. PINEAU, "Low cycle fatigue behavior of Inconel 718 at 298K and 823K", *Met. Trans.*, vol. 8A, pp. 1095-1105, 1977.

[FOU 06a] B. FOURNIER, M. SAUZAY, C. CAES, M. NOBLECOURT, M. MOTTOT, "Analysis of the hysteresis loops of a martensitic steel. Part I: Study of the influence of strain amplitude and temperature under pure fatigue loadings using an enhanced stress partitioning method", *Mat. Science & Eng.*, vol. A 437, pp. 183-196, 2006.

[FOU 06b] B. FOURNIER, M. SAUZAY, C. CAES, M. MOTTOT, M. NOBLECOURT, A. PINEAU, "Analysis of the hysteresis loops of a martensitic steel. Part II: Study of the influence of creep and stress relaxation holding times on cyclic behaviour", *Mat. Science & Eng.*, vol. A 437, pp. 197-211, 2006.

[FOU 07] B. FOURNIER, Fatigue-fluage des aciers martensitiques à 9-12 % Cr: Comportement et endommagement, PhD Thesis, Ecole des Mines de Paris, 2007.

[FOU 08a] B. FOURNIER, M. SAUZAY, C. CAES, M. NOBLECOURT, M. MOTTOT, A. BOUGAULT, V. RABEAU, A. PINEAU, "Creep-fatigue-oxidation interactions in a 9Cr – 1Mo martensitic steel. Part I: Effect of tensile holding period on fatigue lifetime", *Int. J. Fatigue*, vol. 30, pp. 649-662, 2008.

[FOU 08b] B. FOURNIER, M. SAUZAY, C. CAES, M. NOBLECOURT, M. MOTTOT, A. BOUGAULT, V. RABEAU, A. PINEAU, "Creep-fatigue-oxidation interactions in a 9Cr-1Mo martensitic steel. Part II: Effect of compressive holding period on fatigue lifetime", *Int. J. Fatigue*, vol. 30, pp. 663-676, 2008.

[FOU 08c] B. FOURNIER, M. SAUZAY, C. CAES, M. NOBLECOURT, M. MOTTOT, A. BOUGAULT, V. RABEAU, J. MAN, O. GILLIA, P. LEMOINE, A. PINEAU, "Creep-Fatigue-Oxidation interactions in a 9Cr – 1Mo martensitic steel. Part III: Lifetime prediction", *Int. J. Fatigue*, vol. 30, pp. 1797-1812, 2008.

[FOU 09a] B. FOURNIER, M. SAUZAY, C. CAES, M. NOBLECOURT, M. MOTTOT, L. ALLAIS, I. TOURNIE, A. PINEAU, "Creep-fatigue interactions in a 9 Pct Cr-1Mo martensitic steel: Part I. mechanical test results", *Metall. and Mater. Trans. A*, vol. 40A, pp. 321-329, 2009.

[FOU 09b] B. FOURNIER, M. SAUZAY, F. BARCELO, E. RAUCH, A. RENAULT, T. COZZIKA, L. DUPUY, A. PINEAU, "Creep-fatigue interactions in a 9 Pct Cr-1Mo martensitic steel: Part II. Microstructural observations", *Metall. And Mater. Trans. A*, vol 40A, pp. 330-341, 2009.

[GAF 05] V. GAFFARD, Experimental study and modelling of high temperature creep flow and damage behaviour of 9Cr1Mo-NbV steels and weldments, PhD Thesis, Ecole des Mines de Paris, 2005.

[GAT 77] R.S. GATES, "Grain boundary sliding in 316 stainless steel, Part II. The creep strain stress dependencies", *Mat. Sci. & Eng.*, vol. 27, pp. 115, 1977.

[GAY 88] J. GAYDA, T.P. GABB, R.V. MINER, "Isothermal and bithermal thermomechanical fatigue behavior of a Ni NiCoCrAlY-coated single crystal superalloy", in: D.N. DUHL, G. MAURER, S. ANTOLOVICH, C. LUND and S. REICHMAN (eds), *Superalloys 1988*, AIME, Warrendale, PA, pp. 575-584, 1988.

[GEL 86] M. GELL, D.N. DUHL, "Development of single crystal superalloy turbine blades, advanced high-temperature alloys: processing and properties", in S.M ALLEN, R.M. PELLOUX and R. WIDMER (eds), *Proceedings of the Nicholas J. Grant Symposium*, ASMI, Materials Park, Ohio, pp. 41-49, 1986.

[GHO 92] H. GHONEM, D. ZHENG, "Depth of intergranular oxygen diffusion during environment-dependent fatigue crack growth in alloy 716", *Mat. Sci. Eng.*, vol. 150, pp. 151-160, 1992.

[HAL 78] G.R. HALFORD, "A recommended procedure for extrapolating strain fatigue curves beyond one million cycles to failure", prepared for the *Working Group on Creep-Fatigue of the ASME*, March 1978.

[HAL 80] R. HALES, HALES R., "A quantitative metallographical assessment of structural degradation of type 316 stainless steel during creep-fatigue", *Fat. Eng. Mater. Struct.*, vol. 3, pp. 339-356, 1980.

[HAL 91] R. HALES, R.A. AINSWORTH, "Multiaxial creep-fatigue rules", 11th *Int. Conf. on Structural Mechanics in Reactor Technology*, Post Seminar 5, 8th Int. Seminar on Inelastic Analysis, Fracture and Prediction, Tokyo, pp. 125-138, 1991.

[HAY 82] R. HALES, W.C. HAYES, "On the mechanism of delayed discontinuous plastic flow in an age-hardened nickel alloy", *Acta Met.*, vol. 30, pp. 1295-1301, 1982.

[HEC 93] R.L. HECHT, J.R. WEERTMAN, "Periodic oxide cracking on Fe2.25Cr1Mo produced by high temperature fatigue tests with a compression hold", *Met. Trans.* A, vol. 24 A, pp. 327-333, 1993.

[HER 96] R.W. HERTZBERG, *Deformation and Fracture Mechanics of Engineering Materials*, 4th edition, John Wiley and Sons, New York, 1996.

[HUL 01] D. HULL, D.J. BACON, *Introduction to Dislocations*, Butterworth-Heinemann, Woburn, MA, 2001.

[HUR 86] E.S. HURON, S.D. ANTOLOVICH, "Observations of high temperature tensile and cyclic deformation in a directionally solidified nickel-base superalloy, Phase Boundary Effects on Deformation", in: K.N. SUBRAMANIAN and M.A. IMAM (eds), *Phase Boundary effects on Deformation*, TMS-AIME, Warrendale, PA, p. 185-203, 1986.

[JAC 83] B. JACQUELIN, Amorçage des fissures en fatigue oligocyclique sous chargement multiaxial, PhD Thesis, Ecole des Mines de Paris, 1983.

[KAC 58] L.M. KACHANOV, "Tune of the rupture process under creep conditions", *Izv. Akad. Nauk. SSR, Otd Tekn. Nauk.*, vol. 8, pp. 26, 1958.

[KAN 82] F.A. KANDL, M.W. BROWN, K.J. MILLER, *Biaxial Low Cycle Fatigue of 316 Stainless Steel at Elevated Temperatures*, The Metal Society, London, pp. 203-210, 1982.

[KIM 88] S. KIM, J.R. WEERTMAN, " Investigation of microstructural changes in a ferritic steel caused by high temperature fatigue", *Met. Trans. A*, vol. 19A, pp. 999-1007, 1988.

[KIT 06] H. KITAHARA, U. RINTARO, N. TSUJI, Y. MINAMINO, "Crystallographic features of lath martensite in low carbon steel", *Acta Materialia*, vol. 54, pp. 1279-1288, 2006.

[KLU 07] R.L. KLUEH, A.T. NELSON, " Ferritic/martensitic steels for next-generation reactors", *J. Nucl. Mat.*, vol. 371, pp. 37-52, 2007.

[KOU 82] A.K. KOUL, F.B. PICKERING, "Serrated yielding in Ni-Fe base superalloys at 700°C", *Scripta Met.*, vol. 16, pp 119-124, 1982.

[KRU 87] D.D. KRUEGER, S.D. ANTOLOVICH, R.H. VAN STONE, "Effects of grain size and precipitate size on the fatigue crack growth-behavior of alloy 718 at 427°C", *Met. Trans.*, vol. 18A, pp. 1431-1439, 1987.

[LAL 79] C. LALL, S. CHIN, D.P. POPE, "The orientation and temperature dependence of the yield stress of Ni3 (Al,Nb) single crystals", *Met. Trans.*, vol 10A, p. 1323-1332, 1979.

[LAW 85] B. LAWLESS, S.D. ANTOLOVICH, C. BATHIAS, B. BOURSIER, "The effect of microstructure on the fatigue crack propagation and overload behavior of Waspaloy at room temperature", in: J.M. WELLS and J.D. LANDES (eds), *Fracture: Interactions of Microstructure, Mechanisms and Mechanics*, TMS-AIME, Warrendale, PA, pp. 285-301, 1985.

[LEM 85] J. LEMAITRE, J.L. CHABOCHE, *Mécanique des Matériaux Solides*, Dunod, Paris, 1985.

[LER 82] B.A. LERCH, Microstructural effects on the room and elevated temperature low cycle fatigue behavior of Waspaloy, NASA report CR-165497, 1982.

[LER 83a] B.A. LERCH, Das Raum-und Hochtemperatur-Ermuedungsverhalten der Nickelbasislegierung Nimonic 80A (The room and high temperature fatigue behavior of the nickel base alloy Nimonic 80A), PhD dissertation at University of Stuttgart, 1983.

[LER 83b] B.A. LERCH, V. GEROLD, "Deformation of multi-phase and particle containing materials", in: J.B. BILDE-SORENSEN, N. HANSEN, A. HORSEWELL, T. LEFFERS and H. LILHOLT (eds), *Deformation of Multi-phase and Particle Containing Materials: Proceedings of the 4th Riso International Symposium on Metallurgy and Materials Science*, Risv National Laboratory, Roskilde, Denmark, pp. 375-380, 1983.

[LER 84] B.A. LERCH, N. JAYARAMN, S.D. ANTOLOVICH, "A study of fatigue damage mechanisms in Waspaloy from 25 to 800°C", *Mat. Sci. & Eng.*, vol. 66, pp. 151-166, 1984.

[LER 85a] B.A. LERCH, V. GEROLD, "Room temperature deformation mechanisms in Nimonic 80A", *Acta. Met.*, vol. 33, pp. 1709-1717, 1985.

[LER 87] B.A. LERCH, V. GEROLD, "Cyclic hardening mechanisms in Nimonic 80A", *Met. Trans.*, vol. 18A, pp. 2135-2141, 1987.

[LER 90] B.A. LERCH, S.D. ANTOLOVICH, "Fatigue crack propagation behavior of a single crystalline superalloy", *Met. Trans.*, vol. 21A, pp. 2169-2177, 1990.

[LEV 79] C. LEVAILLANT, B. REZGUI, A. PINEAU, "Effects of environment and hold times on high temperature low cycle fatigue behaviour of 316L stainless steel", *Third Int. Conf. on Mechanical Behaviour of Materials*, vol. 2, pp. 163-172, 1979.

[LEV 82] C. LEVAILLANT, A. PINEAU, "Assessment of high-temperature low-cycle fatigue of austenitic stainless steels by using intergranular damage as correlating parameter", in: C. AMZALLAG, B.N. LEIS, P. RABBE (eds), *Low-Cycle Fatigue and Life Prediction*, ASTM STP 770, American Society for Testing and Materials, pp. 169-193, 1982.

[LEV 84] C. LEVAILLANT, Approche métallographique de l'endommagement d'aciers inoxydables austénitiques sollicités en fatigue plastique ou en fluage : description et interprétation physique des interactions fatigue-fluage-oxydation, PhD Thesis, University de Technologie de Compiègne, 1984.

[LEV 88] C. LEVAILLANT, J. GRATTIER, M. MOTTOT, A. PINEAU, "Creep and creep-fatigue intergranular damage in austenitic stainless steels: Discussion of the creep-dominated regime", in: H.D. SOLOMON, G.R. HALFORD, L.R. KAISANT and B.N. LEIS (eds), *Low-Cycle Fatigue, ASTM STP 942*, ASTM, Philadelphia, pp. 414-437, 1988.

[LLO 75] D.J. LLOYD, D.W. CHUNG, M.C. CHATURVEDI, "Serrated yielding in a superalloy (40Co-38Ni-17Cr-5Ti)", *Acta Metall.*, vol. 23, pp. 93-100, 1975.

[MAI 77] P.S. MAIYA, S. MAJUMDAR, " Elevated temperature low cycle fatigue behaviour of different heats of type 304 stainless steel", *Met. Trans.*, vol. 8A, pp. 1651, 1977.

[MAN 53] S.S. MANSON, Behavior of materials under conditions of thermal stress, NASA TN 2933, 1953.

[MAN 65] S.S. MANSON, "Fatigue: A complex subject - some simple approximations", *Exp. Mech.*, vol. 5, pp. 193-226, 1965.

[MAN 71] S.S. MANSON, G.R. HALFORD, M.H. HIRSCHBERG, "Creep-fatigue analysis by strain-range partitioning", *Symposium on Design for Elevated Temperature Environment*, ASME, pp. 12-28, 1971.

[MAN 06] S.S. MANSON, G.R. HALFORD, *Fatigue and Durability of Structural Materials*, ASM International, Materials Park, Ohio, p. 262, 2006.

[McL 62] D. MC LEAN, *Mechanical Properties of Metals*, John Wiley and Sons, New York, 1962.

[McL 78] D. MC LEAN, A. PINEAU "Grain boundary sliding as a correlating concept for fatigue hold-times", *Metal Science*, vol. 12, pp. 313-316, 1978.

[MER 74] H.F. MERRICK, "The low cycle fatigue of three wrought nickel base alloys", *Met. Trans.*, vol. 5, pp. 891-897, 1974.

[MIL 87] W.W. MILLIGAN, S.D. ANTOLOVICH, "Yielding and deformation behavior of the single crystal superalloy PWA 1480", *Met. Trans.*, vol. 18A, pp. 85-95, 1987.

[MIL 88] W.W. MILLIGAN, Deformation modeling and constitutive modeling for anisotropic superalloys, Thesis, Georgia Institute of Technology, 1988.

[MIL 93] M.P. MILLER, D.L. MC DOWELL, R.L.T. OEHMKE, S.D. ANTOLOVICH, "A life prediction model for thermomechanical fatigue based on microcrack propagation, thermomechanical fatigue behavior of materials", *ASTM STP 1186*, ASTM, pp. 35-49, 1993.

[MIN 82] R.V. MINER, J. GAYDA, R.D. MAIER, "Fatigue and creep-fatigue deformation of several nickel-base superalloys at 605°C", *Met. Trans.*, vol. 13A, pp. 1755-1765, 1982.

[MOL 97] R.MOLINS, G. HOCHSTETTER, J.C. CHASSAIGNE, E. ANDRIEU, "Oxidation effects on the fatigue crack growth behavior of alloy 718 at high temperature", *Acta Mater.*, vol. 45, pp. 663-674, 1997.

[MOT 82] M. MOTTO, P. PETREQUIN, C. AMZALLAG, P. RABBE, J. GRATTIER, S.H. MASSON, "Behavior in fatigue relaxation of a high creep resistant type 316L stainless steel", *Low Cycle Fatigue and Life Prediction, ASTM STP 770*, pp. 152-168, 1982.

[MUR 94] T. MURA, "A theory of fatigue crack initiation", *Mat. Sci. & Eng. A,* 176, pp. 61-70, 1994.

[NAH 77] H. NAHM, J. MOTEFF, D.R. DIERCKS, "Substructural development during low cycle fatigue of AISI 304 stainless steel at 649°C", *Acta Met.*, vol. 25, pp. 107-116, 1977.

[NAH 81] H. NAHM, J. MOTEFF, "Characterization of fatigue substructure of Incoloy Alloy 800 at elevated temperature", *Met. Trans. A*, vol. 12A, pp. 1011-1025, 1981.

[PAU 69] D.F. PAULONIS, J.M. OBLAK, D.S. DUVALL, "Precipitation in nickel-base alloy 718", *Trans. ASM 62*, pp. 611-622, 1969.

[PED 82] J.P. PEDRON, A. PINEAU, "The effect of microstructure and environment on the crack growth behavior of Inconel 718 alloy at 650°C under fatigue, creep and combined loading", *Mat. Sci. & Eng.*, vol. 56, pp. 143-156, 1982.

[PIN 10] A. PINEAU, "Low-cycle fatigue", in: C. BATHIAS and A. PINEAU (eds), *Fatigue of Materials and Structures: Fundamentals*, ISTE-Wiley, London-New York, pp. 113-173, 2010.

[PIN 10] A. PINEAU, "Low-cycle fatigue" in C. BATHIAS and A. PINEAU, *Fatigue of Materials and Structures: Fundamentals*, ISTE, London, John Wiley & Sons, New York, 2010.

[PIN 76] A. PINEAU, "Influence of uniaxial stress on the morphology of coherent precipitates during coarsening-elastic energy considerations", *Acta Met.*, vol. 24, pp. 559-564, 1976.

[PIN 80] A. PINEAU, "Influence of micromechanisms of cyclic deformation at elevated temperature on fatigue behaviour", in: R.M. PELLOUX and N.S. STOLOFF (eds), *Creep-Fatigue-Environment Interactions*, ASME Publications, New York, 1980.

[PIN 83] A. PINEAU, "High temperature fatigue behaviour of engineering materials in relation to microstructure", in R.P. SKELTON (ed), *Fatigue at high temperature*, Applied Science Publishers, New York, p. 305-364, 1983.

[PIN 83a] A. PINEAU, "Intergranular creep-fatigue crack growth in Ni-base alloys", Chapter 10, *Flow and Fracture at Elevated Temperatures*, ASME, pp. 317-348, 1983.

[PIN 83b] A. PINEAU, "High temperature fatigue: creep-fatigue-oxidation interaction in relation to microstructure", in: L.H. LARSSON (ed) , Chapter 16, *Subcritical Crack Growth Due to Fatigue, Stress Corrosion and Creep*, Elsevier Applied Science Publishers, pp. 483-530, 1983.

[PIN 89a] A. PINEAU, "Mechanisms of creep-fatigue interactions", in: C. MOURA-BRANCO and L. GUERRA ROSA (eds), *Advances in Fatigue Science and Technology*, Kluwer Academic Press, Dordrecht, Part I, pp. 283-311, 1989.

[PIN 89b] A. PINEAU, "Elevated temperature life prediction methods", in: C. MOURA-BRANCO and L. GUERRA ROSA (eds), *Advances in Fatigue Science and Technology*, Kluwer Academic Press, Dordrecht, Part I, pp. 313-338, 1989.

[PIN 96a] A. PINEAU, "Defect assessment procedures in the creep range", in: C MOURA-BRANCO, R. RITCHIE and V. SKLENICKA (eds), *Mechanical Behaviour of Materials at High Temperature*, Kluwer Academic Publishers, London, Part I, pp. 59-82, 1996.

[PIN 96b] A. PINEAU, "Fatigue and creep-fatigue behaviour of Ni base superalloys : microstructural and environmental effects", in: C. MOURA-BRANCO, R. RITCHIE and V. SKLENICKA (eds), *Mechanical Behaviour of Materials at High Temperature*, Kluwer Academic Publishers, London, Part I, pp135-1584, 1996.

[REE 99] R.C. REED, N. MATAN, D.C. COX, M.A. RIST, C.M.F. RAE, "Creep of CMSX-4 superalloy single crystals: effects of rafting at high temperature", *Acta Mater*, vol. 47, pp. 3367-3381, 1999.

[REM 87] L. REMY, F. REZAL-ARIA, R. DANZER, W. HOFFELNER, "Evaluation of life prediction methods in high temperature fatigue", *Low Cycle Fatigue, ASTM-STP 942*, pp. 115, 1987.

[REU 83a] J. REUCHET, L. REMY, "Fatigue oxidation interaction in a superalloy-application to olife prediction in high temperature low cycle fatigue", *Met. Trans.*, vol. 14A, pp. 141-149, 1983.

[REU 83b] J. REUCHET, L. REMY, "High temperature low cycle fatigue of Mar-M 509 superalloy II- the influence of oxidation at high temperature", *Mat. Sci. Eng.*, vol. 58, pp. 32-43, 1983.

[REZ 82] B. REZGUI, Interaction fatigue-fluage. Effet d'un temps de maintien de traction sur la résistance à la fatigue oligocyclique à 600°C d'un acier Z2CND 17-13 (AISI 316L), PhD Thesis, University of Paris XI, 1982.

[ROB 71] E.L. ROBINSON, " Effect of temperature variation on the long time. Strength of steels", *Trans. ASME*, vol. 74, pp. 777-781, 1952.

[SAL 88] J.F. SALTSMAN, G.R. HALFORD, "An update on the total strain version of srp, Low cycle fatigue-directions for the future", in: H.D. SOLOMON, G.R. HALFORD, L.R. KAISAND and B.N. LEIS (dir.), *ASTM STP 942*, Philadelphia, PA, pp. 329-341, 1988.

[SAN 81] T.H. SANDERS, R.E. FRISHMUTH, G.T. EMBLEY, "Temperature dependent deformation mechanisms of alloy 718 in low cycle fatigue", *Met. Trans.*, vol. 12A, pp. 1003-1010, 1981.

[SAX 75] A. SAXENA, S.D. ANTOLOVICH, "Low cycle fatigue, fatigue crack propagation, and substructures in a series of polycrystalline Cu-Al alloys", *Met. Trans. 6A*, Volume 6A, pp. 1809-1828, 1975.

[SCH 79] B.V.D. SCHAAF, M.I. DE VRIES, J.D. ELEN, Effect of Neutron Irradiation on Sequential Creep-fatigue Interaction of 18Cr-11Ni Stainless Steel Plate and Welded Joints at 823K-ECN74 Report, Netherlands Energy Research Foundation, 1979.

[SHA 84] D.M. SHAH, D.N. DUHL, "The effect of orientation, temperature, and gamma prime size on the yield strength of a single crystal nickel base superalloy" in: M. GELL, C.S. KORTOVICH, R.H. BRICKNELL, W.B. KENT AND J.F. RADAVICH (eds), *Superalloys 1984*, Met. Society AIME, Pittsburgh, PA, pp. 105-114, 1984.

[SHY 05] A. SHYAM, W.W. MILLIGAN, "A model for slip irreversibility, and its effect on the fatigue crack propagation threshold in a nickel-base superalloy", *Acta Mater.*, vol. 53, pp. 835-844, 2005.

[SIM 72] C.T. SIMS, W.C. HAGEL, *The Superalloys*, John Wiley and Sons, New York, 1972.

[SKE 93] R.P. SKELTON, "Damage factors during high temperature fatigue crack growth", in: R.A. AINSWORTH and R.P. SKELTON (eds), *Behavior of Defects at High Temperatures*, Mechanical Eng. Publications, London, pp. 191-218, 1993.

[SPE 73] D.A. SPERA, "Fatigue at elevated temperature", *ASTM STP 520*, pp. 643, 1973.

[SPI 99] S. SPIGARELLI, E. CERRI, P. BIANCHI, E. EBANGELISTA, "Interpretation of creep behaviour of 9Cr-Mo-Nb-V-N (T91) steel using threshold stress concept", *Mat. Sci. & Technol.*, vol. 15, pp. 1143-1440, 1999.

[SPI 04] M.W. SPINDLER, "The multiaxial creep ductility of austenitic stainless steels", *Fat. Fract. Eng. Mater. Struct.*, vol. 27, pp. 273-281, 2004.

[STO 78] R.E. STOLTZ, A.G. PINEAU, "Dislocation precipitate interaction and cyclic stress strain behavior of a γ' strengthened superalloy", *Mat. Sci. & Eng.*, vol. 34, pp. 275-284, 1978.

[SUN 98] W.R. SUN, S.R. GAO, J.H. LEE, N.K. PARK, Y.S. YOO, S.J. CHOE, Z.Q. HU, "Effects of phosphorus on the δ -Ni$_3$Nb phase precipitation and the stress rupture properties in alloy 718", *Mat. Sci. Eng.*, vol. A247, pp. 173-179, 1998.

[SUP 08] Initial acts of 1968: International Symposium on Structural Stability in Superalloys, vol. I and II, AIME, New York, PA, 1968. Now published as Superalloys 2008 .

[TAI 62] S. TAIRA, "Lifetime of structures subjected to varying load and temperature", in: N.J. Hoff (ed), *Creep in Structures*, Academic Press, New York, 1962.

[TAK 73] S. TAKEUCHI, E. KURAMOTO, "Temperature and orientation dependence of the yield stress in Ni$_3$Ga single crystals", *Acta. Met.*, vol. 21, pp. 415-425, 1973.

[TAN 81] K. TANAKA, T. MURA, "A dislocation model for fatigue crack initiation", *J. Appl. Mech.*, vol. 48, pp. 97-103, 1981.

[TIE 89] J.K. TIEN, T. CAUFIELD, *Superalloys, Supercomposites and Superceramics*, Academic Press, New York, 1989.

[TOM 68] B. TOMKINS, "Fatigue crack propagation – An analysis", *Philosophical Magazine*, vol. 18, pp. 1041-1066, 1968.

[UDO 73] T. UDOGUCHI, Y. ASADA, I. ICHINO, "A frequency interpretation of hold time experiments on high temperature low-cycle fatigue of steels for LMFBR", *Int. Conf. on Creep and Fatigue in Elevated Temperature Applications*, 1, Philadelphia, PA, pp. 211, 1973.

[UKA 07] S. UKAI, A. OHTSUKA, "Low cycle fatigue properties of ODS ferritic-martensitic steels at high temperature", *J. Nucl. Mat.*, 367-370, pp. 234-238, 2007.

[VAS 04] A.K. VASUDEVAN, K. SADANANDA, "Analysis of vacuum FCG results and implications", *International Conference on Fatigue Damage of Structural Materials V*, Hyannis, MA, September 19-24, 2004.

[VIS 89] R. VISWANATHAN, *Damage Mechanisms and Life Assessment of High Temperature Components*, ASM International, Materials Park, Ohio, 1989.

[WAR 81] J. WAREING, "Creep-fatigue behaviour of four casts of type 316 stainless steel", *Fat. Eng Mater. & Struct.*, 4, 131-145, 1981.

[WEI 72] J. WEISS, R. STICKLER, "Phase instabilities during high temperature of 316 austenitic stainless steel", *Met. Trans.*, vol. 3, pp. 851-866, 1972.

[WEI 92] J. WEISS, Endommagement en viscoplasticité cyclique sous chargement multiaxial à haute température d'un acier inoxydable austénitique, PhD Thesis, Ecole des Mines de Paris, 1992.

[WEI 93] J. WEISS, A. PINEAU, "Fatigue and creep fatigue damage of austenitic stainless steels under multiaxial loading", Met. Trans., vol. 24A, pp. 2247-2261, 1993.

[WEL 64] C.H. WELLS, C.P. SULLIVAN, "The low cycle fatigue characteristics of a nickel base superalloy at room temperature", Trans. ASM, vol. 57, pp. 841-855, 1964.

[WEL 65] C.H. WELLS, C.P. SULLIVAN, "Low cycle fatigue damage of Udimet 700 at 1400F", Trans. ASM, vol. 58, pp. 391-402, 1965.

[WEL 67] C.H. WELLS, C.P. SULLIVAN, "The effect of temperature on the low cycle fatigue behavior of Udimet 700", Trans. ASM, vol. 60, pp. 217-222, 1967.

[WÖH 1860] A. WÖHLER, "Über die Festigkeitsversuche mit Eisen und Stahl", Zeitschrift für Bauwesen, vol. 10, pp. 583, 1860.

[WRI 04] P.K. WRIGHT, M. JAIN, D. CAMERON, "High cycle fatigue in a single crystal superalloy: time dependence at elevated temperature", in: K.A. GREEN, T.M. POLLOCK, H. HARADA, T.E. HOWSON, R.C. REED, J.J. SCHIRRA, and S. WALSTON (eds), Superalloys 2004, AIME, Warrendale, PA, pp. 657-666, 2004.

[XIA 06] L. XIAO, D.L. CHEN, M.C. CHATURVEDI, "Effect of boron on fatigue crack growth behavior in superalloy IN 718 at RT and 650C", Mat. Sci. Eng., vol. A428, pp. 1-11, 2006.

[XIE 07] X.S. XIE, J.X. DONG, M.C. ZHANG, "Research and development of Inconel 718 type superalloy", Material Science Forum, vol. 539-543, pp. 262-269, 2007.

[YAM 80] K. YAMAGUCHI, K. KANAZAWA, "Influence of grain size on the low-cycle fatigue lives of austenitic stainless steels at high temperatures", Met. Trans. A, vol. 11A, pp. 1692-1699, 1980.

[YOS 86] M. YOSHIDA, C. LEVAILLANT, A. PINEAU, "Metallographic measurement of creep intergranular damage and creep strain, Influence of stress state on critical damage at failure in an austenitic stainless steel", Int. Conf. on Creep, Tokyo, Japan, pp. 327-332, April 14-18, 1986.

[YOS 90] M. YOSHIDA, C. LEVAILLANT, R. PIQUES, A. PINEAU, "Quantitative study of intergranular damage in an austenitic stainless steel on smooth and notched bars", in: P. BENSUSSAN (ed), High Temperature Fracture Mechanisms and Mechanics, Mech. Eng. Publications, London, pp. 3-21, 1990.

Chapter 2

Analysis of Elasto-plastic Strains and Stresses Near Notches Subjected to Monotonic and Cyclic Multiaxial Loading Paths

2.1. Introduction

Increasing demands concerning the optimization of machine components and structures requires methods for predicting the fatigue lives of various complex mechanical engineering components that are subjected to fluctuating load histories. Any fatigue life prediction procedure consists of three main areas that are used to analyze and input data into the calculation procedure:

– material properties;

– loading/stress history; and

– cyclic stress-strain analysis and fatigue damage evaluation.

The material and loading history inputs are similar in most methods; however, the local inelastic stress-strain analyses and the fatigue damage calculation methods may differ depending on the general philosophy used by an analyst. The "information path" used for predicting a fatigue life is shown in Figure 2.1. The loading, geometry and the material stress-strain characteristic need to be combined in order to generate stress-strain response at the critical location, such as the tip in a notched body.

Chapter written by Gregory GLINKA.

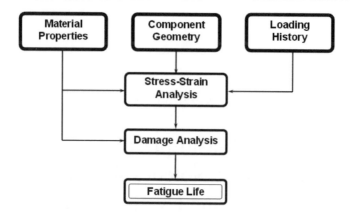

Figure 2.1. *General procedure for fatigue analysis*

Historically, the fatigue testing and the development of predictive models have focused mainly on the uniaxial loading case because it has been applicable to a variety of practical problems and it is the simplest type of fatigue test to conduct. However, an increasing percentage of fatigue design problems involve stress histories induced by fluctuating multiaxial loads. Problems involving multiaxial fatigue cannot be analyzed with models that have been developed for uniaxial fatigue cases, so methods developed for multiaxial loading must be used.

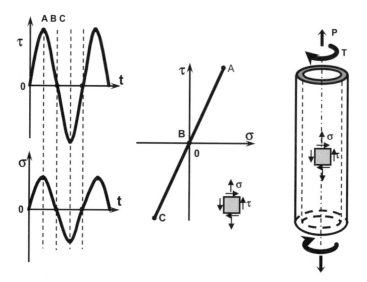

Figure 2.2. *An example of a proportional loading path. The ratio of the stress components remains constant for the entire load history*

Multiaxial fatigue cases arise when more than one mode of loading is applied to a structure at the same time. For example, axles and shafts may experience combined torsion and bending loads. If the various components of cyclic loading are in phase and change proportionally with each other, the loading is termed *proportional* (see Figure 2.2).

If the loading causes the directions of principal stresses and the ratio of principal stress magnitudes to change with each applied loading increment, the loading is termed *non-proportional* (see Figure 2.3). In general, the traditional uniaxial fatigue damage parameters based on the range of one particular normal stress or strain component are not suitable for predicting fatigue caused by multiaxial loading.

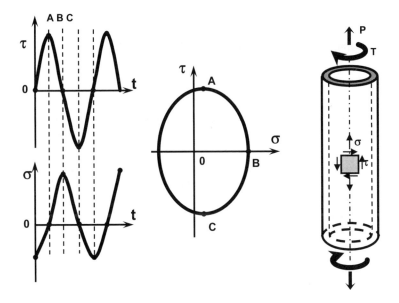

Figure 2.3. *An example of a non-proportional loading path. The ratio of the principal stresses changes because the two components of the loading history are not in phase*

Fatigue durability and strength analysis of machine components and structures subjected to multiaxial cyclic loads also requires determination of elasto-plastic strains and the accumulated fatigue damage at the point where the highest stress concentration occurs. This requires a model that can efficiently simulate the notch-tip stress-strain histories caused by externally applied cyclic loads and fatigue parameters that can account for the damage caused by multiaxial stress states. The resulting local stress-strain histories are subsequently used to determine the amount of fatigue damage at the notch and to estimate the fatigue life of the component.

The main focus of this chapter is on the presentation and critical analysis of some contemporary approaches to multiaxial fatigue in general and multiaxial elastic-plastic notch stress-strain analyses in particular.

2.2. Multiaxial fatigue parameters

Numerous methods have been proposed to deal with the problem of multiaxial fatigue. The majority of multiaxial fatigue theories fall into two broad categories; those that quantify the fatigue damage with using equivalent parameters, such as the equivalent stress, strain or energy; and those that relate fatigue damage to a particular failure plane, or *critical plane*.

2.2.1. *Equivalent parameter methods*

2.2.1.1. *Equivalent stress and strain*

Equivalent stress and strain methods were the first methods used to attempt to reconcile the problems of multiaxial fatigue. They were a logical first step, as they were extensions of the existing static failure criteria that had been developed for predicting failures of structures subject to multiaxial static loads. These methods essentially reduce a 3D stress or strain state to an equivalent scalar stress or strain parameter. The equivalent stress or strain may be found by way of the maximum principal stress, the maximum shear (Tresca criterion), or the maximum distortion strain energy density (von Mises or octahedral shear stress criterion). The calculated equivalent parameter is then compared to either the stress or the strain based uniaxial fatigue properties.

Equivalent stress- and strain-based methods have the advantage of being relatively simple to implement for proportional loading cases. However, despite the convenience delivered by these methods, they are also subject to some serious limitations. Tipton and Nelson [TIP 89] and Tipton and Fash [TIP 88] conducted comparisons of various stress- and strain-based approaches. They used experimental data from a round robin test program that was coordinated by the Society of Automotive Engineers in the late 1980s [KUR 88]. All the equivalent stress and strain approaches were found to be non-conservative for non-proportional loading. This was due mainly to the lack of the load path dependence inherent in this type of fatigue parameter. Also, equivalent stress and strain theories do not predict mean stress effects on the fatigue life; it is well-known that a tensile mean stress, for example, causes a significant decrease in the fatigue life.

The most popular stress-based criteria are extensions of classical failure theories, such as the maximum principal stress, the maximum shear stress or the octahedral shear stress. These stress-based criteria are most often employed in the analysis of high-cycle fatigue problems where very little plasticity occurs. None of these theories, however, are sufficient to correlate various loading modes in situations where cyclic plasticity is involved.

One of the most well-known and complete equivalent stress methods was presented by Sines [SIN 59]. Sines examined many sets of experimental fatigue data and concluded that at given fatigue life a linear relationship exists between the effect of static stresses and the allowable alternating stresses. He subsequently modified the equivalent octahedral shear stress theory to account for his findings. The model Sines proposed was of the following form:

$$\frac{1}{3}\sqrt{(p_1 - p_2)^2 + (p_2 - p_3)^2 + (p_1 - p_3)^2} + \alpha(S_x + S_y + S_z) \le A \qquad [2.1]$$

where $p_1, p_2,$ and p_3 are the amplitudes of the alternating principal stresses and S_x, S_y and S_z are orthogonal static stresses. α and A are material constants that can be determined by using the results of two different types of loading that result in the same fatigue life.

Sines suggested using the results for loading cases for $R = 0$ and $R = -1$. He showed that this criterion worked well for the fatigue data that were available. This model accounted for the effects of both cyclic and hydrostatic static mean stress on fatigue life. It was intended to be used to design components for infinite fatigue lives ($>10^7$ cycles) by allowing the designer to calculate a stress level for the expected cyclic loading conditions and then compare it to a fatigue endurance limit.

This criterion has limited usefulness as a fatigue life prediction tool since it does not provide a convenient method for relating the applied cyclic stresses to the expected fatigue life without conducting many experiments for two distinct loading modes at a variety of fatigue lives. While the limited number of results support the application of this model to non-proportional loading cases, Sines emphasizes that yielding must be avoided if this criterion is to be used. This restriction prevents its use for any loading that may cause plastic strains at the notch tip. The linear nature of the mean stress effects does not hold when plasticity occurs in the critical region.

The most popular strain-based theories are strain versions of the classical failure theories mentioned above, i.e. the octahedral strain that has shown some capability of correlating various modes of cyclic loading. Nevertheless, it is often found that strain-only based criteria are not sufficient to model a wide variety of loading modes. In particular, the mean stress effect is not taken into account.

2.2.1.2. *Strain energy density methods*

Strain energy density-based criteria have been proposed and tested by several researchers [ELL 84, GAR 81, RAD 80], but have not been generally accepted. The main criticism is that strain energy density, which is a scalar quantity, is unable to account for the loading history effects. However, the advantage of an energy criterion is that it incorporates both strains and stresses into the fatigue damage parameter.

Energy approaches attempt to relate work, or strain energy density, to fatigue life. The use of strain energy density as a fatigue damage parameter for uniaxial loading is not a new one. Feltner and Morrow [FEL 61] reported that the idea had been considered as early as 1927 by Inglis, who measured the total energy to fracture in rotating bending specimens. The use of the total strain energy density as a fatigue-correlating parameter for combined loadings was refuted in 1961 by Findley *et al.* [FIN 61], who performed critical experiments by using a specimen for which the strain energy density at the critical location was kept constant while the principal stress axes rotated. The net effect was that, while the strain energy level remained constant, the stresses acting on any selected plane in the critical region fluctuated. Since fatigue cracks initiated in the region of constant strain energy, it showed that fluctuation of strain energy was not a valid parameter to be used for multiaxial loading conditions.

Garud [GAR 79] has proposed using the plastic work per load cycle as a fatigue correlating parameter:

$$W_c^P = \int_{cycle} \sigma_{ij} d\varepsilon_{ij}^P \qquad [2.2]$$

In equation [2.2], W_c^P is the plastic work over one load cycle, σ_{ij} are the stress components and ε_{ij}^P are the plastic strain components. The resulting fatigue theory was proposed to be:

$$N_f = A\left(W_c^P\right)^\alpha \qquad [2.3]$$

where N_f is the number of loading cycles to failure and A and α are constants to be determined from fatigue experiments.

Garud used his criterion to compare predicted fatigue lives to actual lives from uniaxial, torsional and out-of-phase multiaxial fatigue tests. He found that the proposed plastic work criterion resulted in conservative predictions for uniaxial loading cases.

While Garud [GAR 79] stated that "using this approach requires only the uniaxial cyclic stress-strain curve and the uniaxial fatigue test results on smooth specimens", he also found it necessary to include an empirically derived "weighting factor" of 0.5 in the plastic work done by applied shear strains in order to make the axial and torsional loading fatigue lives correlate on the same plot.

Jordan et al. [JOR 85] compared fatigue life predictions using this model with experimental data of their own and they found that the results diverge for lives in excess of 2,000 cycles where the plastic strain energies are very small. No method for handling multiaxial mean stress effects or hydrostatic pressure effects on fatigue endurance has been proposed for this theory.

Despite the earlier findings of Findley et al. [FIN 59] concerning the unsuitability of total strain energy density as a multiaxial fatigue damage criterion, Ellyin and Golos [GOL 88] tried to use the total strain energy density per cycle, ΔW^t, as a damage parameter for multiaxial fatigue:

$$\Delta W^t = \kappa N_f^\alpha + C \qquad\qquad [2.4]$$

In equation [2.4], N_f is the number of cycles to failure, and κ, α and C are constants that need to be evaluated for a particular material. Citing previously published works, Findley et al. stated that, "the total strain energy density is shown to be a proper damage parameter for uni-axial stress states" [FIN 59].

Based on this assertion, they extend the theory by deriving strain energy relationships for multiaxial fatigue. Their model was developed to include the effects of mean stress and hydrostatic stresses in the calculation of ΔW^t for cases of proportional and nearly proportional loading. This model cannot be used with non-proportional loading cases because the plastic components of the total energy expression were evaluated using the theory of total deformation plasticity. Application of this method also results in different fatigue life curves for each ratio of the maximum principal strain.

2.2.2. Critical plane methods

Reasonably good correlations of multiaxial fatigue data have been obtained, in general, by using critical plane approaches [FIN 59, KAN 82, MCD 91]. These theories postulate that fatigue cracks initiate and grow on certain planes and that only the shear and normal stresses and strains acting on these planes contribute to the fatigue crack formation and growth. In addition, these theories are supported by various experimental observations of the physical mechanism of the fatigue damage process [BAN 92, FAT 88, KAN 82, LOH 80].

Critical plane models have been proposed by several researchers, such as Findley [FIN 59], McDiarmid [MCD 91], Kandil *et al.* [KAN 82] and Lohr and Ellison [LOH 80]. Most of the criteria are given in the form of expressions involving a combination of the stress [MCD 91] or strain [KAN 82] components associated with the critical plane (see Figure 2.4). In general, the strain criterion can be given in the form of the critical plane shear and normal strain components, as proposed by Kandil *et al.* [KAN 82]:

$$\gamma^* = \frac{\Delta\gamma_{21}}{2} + K\frac{\Delta\varepsilon_{22}}{2}$$ [2.5]

McDiarmid [MCD 91] proposed an analogous stress criterion associated with the critical plane:

$$\tau^* = \frac{\Delta\sigma_{21}}{2} + k\frac{\Delta\sigma_{22}}{2}$$ [2.6]

The notation used in equations [2.5] and [2.6] is shown in Figure 2.4. These parameters are sometimes criticized for the lack of formal correctness from the continuum mechanics viewpoint.

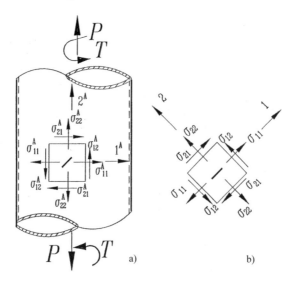

Figure 2.4. *Coordinate system used in the multiaxial fatigue analysis: a) system of 1^A and 2^A coordinates associated with the global axes of the body and applied stress components $\sigma_{ij}{}^A$; b) system of 1 and 2 coordinates associated with the critical plane and stress components σ_{ij} in the critical plane*

The main difficulty concerns the interpretation of these equations, which represent an algebraic sum of the shear and normal strain or stress components acting in the critical plane.

Also, the shear and normal components are weighted using an experimental parameter, K or k, hoped to be constant for a given material [MCD 91]. Fatemi and Kurath [FAT 88] have shown that the weighting parameter K in equation [2.5] may vary with fatigue life. It has been suggested [BAN 92] that in the case of multiaxial fatigue, estimates of lifespan should be based on a combination of both strains and stresses, because they both contribute to cyclic damage.

With this in mind, the effects of stress and strain may be accounted for by applying a strain energy density relation [GLI 95a] formulated on the critical plane that is analogous to equations [2.5] and [2.6]:

$$W^* = \frac{\Delta\gamma_{21}}{2}\frac{\Delta\sigma_{21}}{2} + \frac{\Delta\varepsilon_{22}}{2}\frac{\Delta\sigma_{22}}{2} \qquad [2.7]$$

The novelty of the fatigue strain energy parameter W^* lies in the fact that it represents only that fraction of the strain energy contributed by the stresses and strains on the critical plane, not the overall strain energy density at the point. Parameter W^* is load path-dependent, contrary to the criteria based on the overall strain energy density contributed by all stress and strain components.

In addition, equation [2.7] is acceptable from the point of view of the formalism of continuum mechanics because the energy components $\Delta\sigma_{21}\Delta\gamma_{21}$ and $\Delta\sigma_{22}\Delta\varepsilon_{22}$ are scalars and can be added algebraically. Thus, parameter W^* includes all of the important features often mentioned in the literature:

– the parameter is associated with a favorable or critical plane;

– it is load path-dependent;

– it is formally correct from the continuum mechanics viewpoint; and

– it does not include any empirical fitting parameters.

However, in the form expressed by equation [2.7], parameter W^* cannot account for the effects of mean stress. Further modifications are therefore needed.

2.2.3. *Mean stress effects in multiaxial fatigue*

Extensions of critical plane strain and energy criteria accounting for mean stress effects have been proposed by many researchers. These extensions are based on a few important observations [FIN 53, SOC 89]. First, the alternating shear stress and strain on the critical plane are the primary cause of fatigue. Second, the mean or maximum values of the normal and shear stress and strain components on the critical plane are important secondary contributors to fatigue damage. Consequently, several versions of the critical plane strain criterion that account for the effect of the normal mean stress have been proposed. The criteria used by Fatemi and Kurath [FAT 88] and Socie et al. [SOC 89] show some capability of correlating multiaxial fatigue data obtained at various mean load levels:

$$\gamma^* = \frac{\Delta\gamma_{21}}{2}\left(1 + k\frac{\sigma_{22}^{max}}{S_y}\right) \qquad [2.8]$$

Equation [2.8] accounts for the mean normal stress effect but ignores the mean shear stress effect. On the other hand, the formally-correct strain energy criteria that account for mean stress effects often ignore the existence of the critical plane. One such criterion has been proposed recently by Ellyin et al. [ELL 88]:

$$W^* = \Delta W^e + \Delta W^p = f\left(N_f\right) \qquad [2.9]$$

It can be shown [MOF 95, SIN 96] that the overall strain energy density does not depend strongly on the load path because it only depends on the current stress state. For this reason equation [2.9] may not show a significant load path dependency despite the fact that one of the energy terms is indeed load path dependent. Nevertheless, the combination of plastic work per cycle and the elastic strain energy holds some promise.

A strain energy density criterion associated with the critical plane concept has been also proposed by Chu et al. [CHU 92]. It combines both maximum stresses with the corresponding strain amplitudes on the critical plane in the form analogous to the well-known Smith-Watson-Topper parameter:

$$W^* = 2\sigma_{21}^{max}\gamma_{21}^a + \sigma_{22}^{max}\varepsilon_{22}^a \qquad [2.10]$$

The mean stresses (shear and normal) can be accounted for in equation [2.10], providing that both stress-strain components are alternating. The mean stress effect of constant normal stress in equation [2.10] is null if $\varepsilon_{22}^a = 0$.

Based on the considerations above, a modified strain energy density criterion associated with the critical plane concept has been proposed [GLI 95b] in the following form:

$$W^* = \frac{\Delta\gamma_{21}}{2}\frac{\Delta\sigma_{21}}{2}\left[\frac{1}{1-\sigma_{21}^{max}/\tau_{f'}} + \frac{1}{1-\sigma_{22}^{max}/\sigma_{f'}}\right] \qquad [2.11]$$

The form of equation [2.11] has been determined after considering the results of fatigue experiments carried out under constant amplitude proportional and non-proportional multiaxial loading [BAN 92, FAT 88, KAN 82, LOH 80, MCD 91]. For these results it was concluded that the leading fatigue parameters were the alternating shear stress, $\Delta\sigma_{21}$, and strain, $\Delta\gamma_{21}$ on the critical plane. The maximum normal stress, σ_{22}^{max}, and maximum shear stress, σ_{21}^{max}, are used in the equation to account for the mean stress effect. The normal stress on the critical plane assists in opening the crack and thus accelerates its growth. Similarly, the maximum shear stress effectively helps us to overcome any sliding friction that takes place between the crack surfaces.

The advantages of equation [2.11], when compared with equation [2.7], are that the mean stresses have been taken into account and that only the alternating shear stress and strain are considered for the cyclic effect. This is in agreement with many experimental fatigue observations. Another advantage is that the parameter of equation [2.11] gives the orientation of the critical plane, since its maximum coincides with the plane of maximum shear strain amplitude for the case of proportional loading.

In addition, parameter W^* will predict different lifespans for the case of static compressive and tensile loads superposed on cyclic shear loading. This is in agreement with experimental observations [KOC 85].

2.2.4. Predictive capabilities of multiaxial parameter W*

In order to assess the capability of the fatigue strain energy parameter, W^*, given by equation [2.11], to correlate fatigue lives corresponding to various loading modes, fatigue data obtained under different multiaxial loading conditions and different mean stresses were analyzed and plotted in the form of W^* versus N_f. The set of experimental data was taken from Kurath et al. [KUR 88], who tested thin-walled tubular specimens (see Figure 2.5) made from hot rolled and normalized SAE 1045 steel.

The specimens were thin-walled tubes that were 2.54 mm thick, 210 mm long and had a 25.4 mm internal diameter. The gage length was 33 mm. The specimens were tested under three different loading conditions: tension, torsion, and simultaneous proportional tension and torsion. A constant ratio of applied shear-to-normal strain range $\Delta\gamma_{12}^A/\Delta\varepsilon_{22}^A$ was maintained during each test. This ratio varied from 0 to ∞ from test to test. The number of cycles, N_f, obtained from testing tubular specimens coincided with the creation of a surface crack 1 mm long.

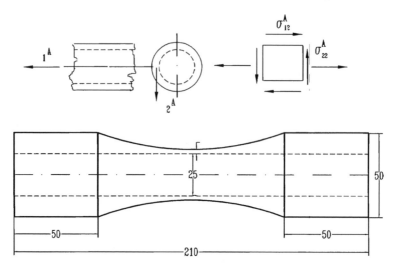

Figure 2.5. *Geometry and dimensions of tubular specimen tested by Kurath et al. [KUR 88]. Coordinate system is shown (dimensions in mm)*

The uniaxial fatigue data obtained from cylindrical specimens 6 mm in diameter tested under constant amplitude strain control were also available in the form of the well known Coffin-Manson equation:

$$\frac{\Delta\varepsilon}{2} = \frac{\sigma_f'}{E}(2N_f)^b + \varepsilon_f'(2N_f)^c \qquad [2.12]$$

First, parameter W^* for the uniaxial tests involving solid cylindrical specimens was determined using equation [2.11]. The plot $W^* - N_f$ for the uniaxial data, denoted by a solid line, is shown in Figure 2.6.

Parameter W^* in equation [2.11] was again plotted against the experimental fatigue life, N_f, obtained from tubular specimens and the results denoted by symbols are also shown in Figure 2.6. It is apparent that all of the data, regardless of the type of loading and mean stress level, collapse onto one line. This can be satisfactorily approximated by the standard curve determined from cylindrical specimens tested under fully reversed uniaxial loading (equation [2.11]).

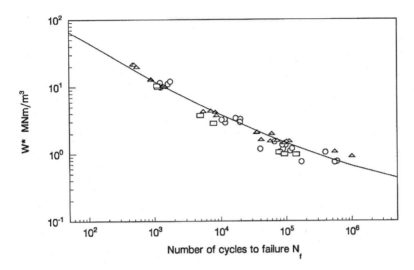

Figure 2.6. *Strain energy density parameter W^* (equation [2.11]) versus experimental fatigue life N_f data for tubular SAE 1045 steel specimens tested under cyclic tension and torsion. (O – mixed loading; Δ – pure torsion; $\square$ – pure tension; —— standard $\Delta\varepsilon$ – N curve)*

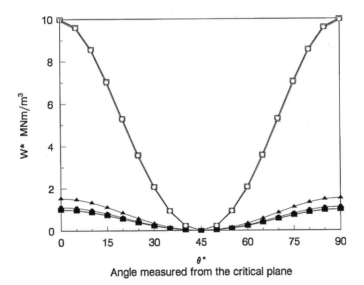

Figure 2.7. *Variation of parameter W^* (equation [2.11]) with orientation of plane for SAE 1045 steel. ($\square$ – pure tension, $N_f = 1,137$; Δ – pure torsion, $N_f = 1,269$; O – mixed loading, $N_f = 1,258$; $\blacksquare$ – pure tension, $N_f = 94,525$; Δ – pure torsion, $N_f = 102,100$; $\bullet$ – mixed loading, $N_f = 115,500$)*

Figure 2.7 shows the variation of parameter $W*$ with orientation of the plane from the maximum shear strain plane for the SAE 1045 steel specimens at various levels and configurations of the applied loading. All of the curves in Figure 2.7 indicate that the plane of maximum damage coincides with the plane of maximum shear strain amplitude.

The second set of experimental data was taken from that reported by Fash [FAS 85]. Constant amplitude load-controlled fatigue tests were conducted on shaft specimens (see Figure 2.8) made of SAE 1045 steel. The cracks initiated at the smaller shoulder radius and the fatigue life were defined as number of cycles, N_f, required to grow a 1 mm long crack on the surface. The specimens were tested under fully-reversed bending, torsion, and simultaneous bending and torsion. The ratio of the bending to torsion moment M_b/M_t was kept constant during each test, resulting in proportional cyclic loading. However, due to plastic yielding at the notch, the stress state at the notch tip did not remain exactly proportional. Therefore, the incremental Neuber approach to the notch-tip stress-strain analysis was adopted.

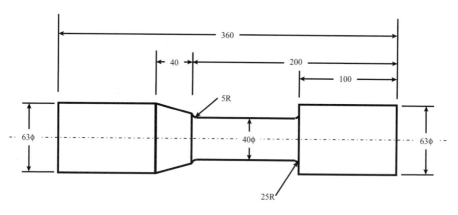

Figure 2.8. *Geometry and dimensions (mm) of the notched shaft specimen [FAS 85] tested under bending and torsion*

The fatigue damage was determined using the fatigue parameter of equation [2.11]. The fatigue life was calculated using the critical plane strain energy versus the fatigue life relationship ($W* - N_f$) obtained from uniaxial strain-controlled tests. A comparison of predicted and experimental lives is shown in Figure 2.9.

The whole procedure discussed above is capable of predicting fatigue lives for a variety of multiaxial loading modes using only the uniaxial cyclic stresses-strain and fatigue strain-life curves. However, the predicted lives tend to be conservative in the high-cycle fatigue regime.

All the data that have been analyzed up to this point were obtained under proportional multiaxial cyclic loading. It is yet to be determined whether or not the methods described above will be equally successful at predicting fatigue lives for non-proportional loading. It also apparent that a practically feasible and efficient method is needed in order to calculate all stress and elasto-plastic strain components at the notch tip induced by complex multiaxial loading histories.

The other family of multiaxial fatigue damage parameters is the group of incremental damage parameters making it possible to estimate fatigue damage for small increments of the stress and strain path.

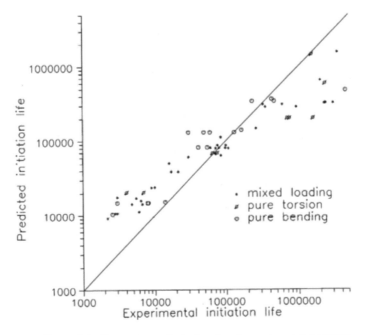

Figure 2.9. *Comparison of calculated and experimental [FAS 85] fatigue lives for the shaft specimens made from SAE 1045 steel*

The fatigue damage evaluation requires integration of the incremental damage parameter along the applied stress-strain path. The advantage of using incremental damage parameters lies in the fact that they do not require identification of stress-strain cycles that are difficult to uniquely define in the case of non-proportional multiaxial cyclic loading histories. Examples of the incremental formulation of fatigue damage parameters can be found in early papers by Ott *et al.* [OTT 87] and Dang [DAN 89].

A thorough review of various multiaxial fatigue damage criteria has been presented by Marquis and Socie in their monograph [SOC 00] and by Kallmeyer in a review paper [KAL 02]. A very original approach to fatigue damage evaluation under multiaxial loading, that can also be classified as the incremental/integral approach to fatigue damage evaluation, has been proposed by Stefanov [STE 97]. The advantage of using the incremental fatigue damage parameter lies in the fact that it does not require us to define and to count load cycles that are difficult to uniquely define in the case of cyclic non-proportional multiaxial loading.

The studies on fatigue damage under multiaxial loadings are still being carried out and none of the parameters discussed above or those analyzed in the references [KAL 02, SOC 00] can be recommended as being the most reliable or appropriate.

2.3. Elasto-plastic notch-tip stress-strain calculation methods

In order to estimate fatigue damage accumulation at the notch tip it is necessary to determine the stress-strain history induced by the external load. The most frequently used methods for calculating the elastic-plastic notch-tip strains and stresses induced by cyclic loads are the Neuber rule [NEU 61], which has been extended to fatigue problems by Topper *et al.* [TOP 69], and the equivalent strain energy density (ESED) method [MOL 81].

An extension of Neuber's rule for multiaxial stress states has been proposed by Hoffman and Seeger [SEE 86] and by Barkey and Socie [BAR 94]. A more general extension of Neuber's rule and the ESED method for multiaxial loading has been proposed by Moftakhar *et al.* [MOF 95]. The approach is based on consideration of the strain energy density at the notch tip and on the observation that the strain energy density at the notch tip can be obtained from the linear elastic stress-strain solution of the notch boundary problem. The method discussed in [MOF 95] is concerned with multiaxial proportional loaded histories.

A similar formulation, appropriate for notched bodies subjected to monotonic non-proportional multiaxial loads, can however be found in reference [SIN 96]. The extension of those models to multiaxial cyclic loading histories is discussed below.

2.3.1. *Uniaxial strain or plane strain states at the notch tip*

If the dimensions and external loads of a body are such that a state of plane stress exists in the body, the stress state at the notch tip is uniaxial (see Figure 2.10).

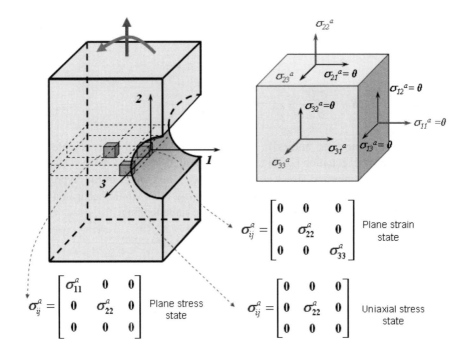

Figure 2.10. *Various stress states in the notch-tip region*

In such a case, four independent relations are required to define one unknown notch-tip stress and three unknown strain components. Similarly, if the notched body is in a state of plane strain (see Figure 2.10), four relations are needed to determine the four unknowns: two non-zero notch-tip strains and two non-zero stress components. Three independent relationships can be obtained from material constitutive equations and only one additional relationship is required.

The additional relationship that is required for the plane stress or the plane strain state at the notch tip is generally defined by either the Neuber rule [NEU 61] or the ESED [MOL 81] relation. Both models relate the fictitious "linear elastic" stresses and strains at the notch tip (σ_{ij}^e, ε_{ij}^e) to the actual elastic-plastic stresses and strains (σ_{ij}^a, ε_{ij}^a), as shown in Figure 2.11.

2.3.1.1. *The Neuber rule*

The Neuber rule [NEU 61] was initially proposed for a notched body loaded in pure shear, but is most often used for notches under tensile or bending loads (see Figure 2.12). It relates the theoretical stress concentration factor, K_t, to the actual stress concentration, K_σ, and the strain concentration, K_ε, as follows:

$$K_t^2 = K_\sigma K_\varepsilon \qquad\qquad [2.13]$$

where:

$$K_t = \frac{\sigma_{22}^e}{\sigma_n}; \ K_\sigma = \frac{\sigma_{22}^N}{\sigma_n}; \ K_\varepsilon = \frac{\varepsilon_{22}^N}{\varepsilon_n}; \ \varepsilon_n = \frac{\sigma_n}{E}$$

In the case of notched bodies in plane stress for which the stress state in the notch tip is uniaxial, equation [2.13] can also be written in another form. This relates the elasto-plastic strain and stress components (ε_{22}^N and σ_{22}^N) to the hypothetical linear elastic notch-tip strain and stress (ε_{22}^e and σ_{22}^e) that would occur in a purely elastic body with the same geometry and under the same loading:

$$\sigma_{22}^e \, \varepsilon_{22}^e = \sigma_{22}^N \, \varepsilon_{22}^N \qquad\qquad [2.14]$$

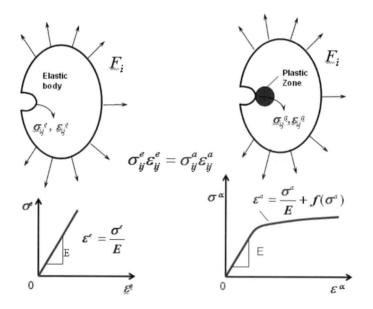

Figure 2.11. *Geometrically identical elastic and elastic-plastic bodies subjected to identical boundary conditions*

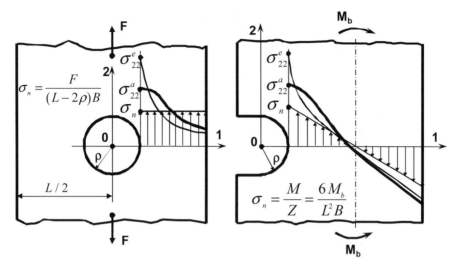

Figure 2.12. *The nominal (σ_n), hypothetical elastic (σ_{22}^e) and actual elastic-plastic (σ_{22}^a) stress distributions near notches for tensile loading (left) and bending load (right)*

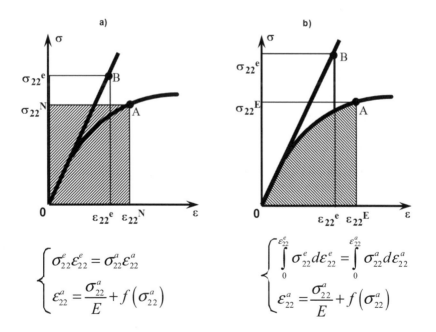

$$\begin{cases} \sigma_{22}^e \varepsilon_{22}^e = \sigma_{22}^a \varepsilon_{22}^a \\ \varepsilon_{22}^a = \dfrac{\sigma_{22}^a}{E} + f\left(\sigma_{22}^a\right) \end{cases} \qquad \begin{cases} \displaystyle\int_0^{\varepsilon_{22}^e} \sigma_{22}^e d\varepsilon_{22}^e = \int_0^{\varepsilon_{22}^a} \sigma_{22}^a d\varepsilon_{22}^a \\ \varepsilon_{22}^a = \dfrac{\sigma_{22}^a}{E} + f\left(\sigma_{22}^a\right) \end{cases}$$

Figure 2.13. *Graphical interpretation: a) Neuber's rule; and b) the equivalent strain energy density method*

Thus, when the notch tip is subjected to a uniaxial stress state (as for a body in plane stress), equation [2.14] represents the equality of the total strain energy density at the notch tip, represented by the rectangles A and B in Figure 2.13a.

The total strain energy density is defined as the sum of the strain energy density and the complementary strain energy density. A relationship similar to equation [2.14] can also be written for notched bodies in plane strain.

The uniaxial constitutive stress-strain curves for engineering materials are most often given in the following general form:

$$\varepsilon = \frac{\sigma}{E} + f(\sigma) \tag{2.15}$$

For a uniaxial constitutive law in the form of equation [2.15], the set of four equations that is necessary to determine all of the stress and strain components at the notch tip of a body in plane stress are given below:

$$\varepsilon_{11}^N = -\nu \frac{\sigma_{22}^N}{E} - \frac{1}{2} f(\sigma_{22}^N)$$

$$\varepsilon_{22}^N = \frac{\sigma_{22}^N}{E} + f(\sigma_{22}^N)$$

$$\varepsilon_{33}^N = -\nu \frac{\sigma_{22}^N}{E} - \frac{1}{2} f(\sigma_{22}^N) \tag{2.16}$$

$$\varepsilon_{22}^e \sigma_{22}^e = \varepsilon_{22}^N \sigma_{22}^N$$

2.3.1.2. *Equivalent strain energy density rule*

The ESED relationship [MOL 81] was initially proposed for a notched body in plane stress. It is written as the equality of the strain energy densities at the notch tip:

$$\int_0^{\varepsilon_{22}^e} \sigma_{22}^e \, d\varepsilon_{22}^e = \int_0^{\varepsilon_{22}^E} \sigma_{22}^E \, d\varepsilon_{22}^E \tag{2.17}$$

Equation [2.17] is a statement of the equality between the strain energy density, W^e, at the notch tip of a linear elastic body and the strain energy density, W^E, at the notch tip of a geometrically identical elastic-plastic body subjected to the same load (see Figure 2.11). This relationship is shown graphically in Figure 2.13b, and represents the equality of the area-under-the-linear-elastic curve and the area-under-the-actual-elastic-plastic $\sigma^a - \varepsilon^a$ material curve.

For the uniaxial stress-strain constitutive law given in the form of equation [2.15], the relationships necessary to determine all of the stress and strain components at the notch tip of a body in plane stress are given in the form of four equations:

$$\varepsilon_{11}^E = -v\frac{\sigma_{22}^E}{E} - \frac{1}{2}f(\sigma_{22}^E)$$

$$\varepsilon_{22}^E = \frac{\sigma_{22}^E}{E} + f(\sigma_{22}^E)$$

$$\varepsilon_{33}^E = -v\frac{\sigma_{22}^E}{E} - \frac{1}{2}f(\sigma_{22}^E)$$

$$\frac{1}{2}\varepsilon_{22}^e\sigma_{22}^e = \int \sigma_{22}^E d\varepsilon_{22}^E$$

[2.18]

The material curve $f(\sigma_{22}^E)$ relating the plastic strain and stress is to be determined experimentally.

2.3.2. Multiaxial stress states

For the case of a general multiaxial loading applied to a notched body, the state of stress near the notch tip is triaxial. At the notch tip, the stress state is biaxial because of the free surface (see Figure 2.10c). Since equilibrium of the element at the notch tip must be maintained, $\sigma_{23} = \sigma_{32}$ and $\varepsilon_{23} = \varepsilon_{32}$.

Therefore, there are seven unknowns at the notch tip; three stress components and four strain components. A set of seven independent equations is required to completely define the stress-strain state at the notch tip. The material constitutive relationships provide four equations, leaving three additional relationships to be established.

2.3.2.1. Proportional loading

2.3.2.1.1. The constitutive equations

The generalized elastic-plastic constitutive relationships are derived from the uniaxial stress-strain curve by using the principles of elasticity and plasticity theory. The simplest analysis uses a general form of Hencky's total deformation equations:

$$\varepsilon_{ij} = \frac{1+v}{E}\,\sigma_{ij} - \frac{v}{E}\,\sigma_{kk}\,\delta_{ij} + \frac{3}{2}\,\frac{\varepsilon_{eq}^{p}}{\sigma_{eq}}\,S_{ij} \qquad\qquad [2.19]$$

where:

$$\sigma_{eq} = \sqrt{\frac{3}{2}\,S_{ij}\,S_{ij}} \quad \varepsilon_{eq}^{p} = \sqrt{\frac{2}{3}\,\varepsilon_{ij}^{p}\,\varepsilon_{ij}^{p}}$$

$$S_{ij} = \sigma_{ij} - \frac{1}{3}\,\sigma_{kk}\,\delta_{ij} \qquad \sigma_{kk} = \sigma_{11} + \sigma_{22} + \sigma_{33}$$

It is also assumed that:

$$\varepsilon_{eq}^{p} = f(\sigma_{eq}) \qquad\qquad [2.20]$$

The function $f(\sigma_{eq})$ is identical to the relationship between stress and plastic strain for a uniaxial tension or compression test.

The case of the very popular Ramberg-Osgood stress-strain material curve is usually written in the form of equation [2.20a]:

$$\varepsilon = \frac{\sigma}{E} + \left(\frac{\sigma}{K}\right)^{\frac{1}{n}} \qquad\qquad [2.20a]$$

The equivalent plastic strain versus equivalent stress can be written in such a case as:

$$\varepsilon_{eq}^{p} = \left(\frac{\sigma_{eq}}{K}\right)^{\frac{1}{n}} \qquad\qquad [2.20b]$$

Therefore, Hencky's equations can be finally written in terms of stresses only:

$$\varepsilon_{ij} = \frac{1+v}{E}\,\sigma_{ij} - \frac{v}{E}\,\sigma_{kk}\,\delta_{ij} + \frac{3}{2}\,\frac{f(\sigma_{eq})}{\sigma_{eq}}\,S_{ij} \qquad\qquad [2.21]$$

Note that the results from Hencky's equations are identical to those of the incremental theory of plasticity (which is presented later) when the deviatoric stresses remain in fixed proportions.

2.3.2.1.2. Generalized Neuber's and ESED rules

Moftakhar *et al.* [MOF 95] generalized both the Neuber rule and the ESED method to address multiaxial notch-tip stress states in bodies subjected to proportional loading.

They proposed that a band exists within which the actual notch-tip strains will always fall. They also found that the lower bound of this band is obtained by using the generalized ESED method based on equation [2.17] and the upper bound is given by Neuber's rule in the form of equation [2.14].

One of the three additional equations required to define the notch-tip stress-strain field is based on the hypothesis that the equations of strain energy density for uniaxial loading should also hold for the case of multiaxial stress states. This is provided that the strain energy density is calculated by accounting for all of the stress and strain components at the notch tip:

$$\sigma_{22}^e \, \varepsilon_{22}^e = \sigma_{22}^N \, \varepsilon_{22}^N \qquad [2.22]$$

On the principal stress axes, equation [2.22] can be reduced to a simple form consisting of the principal stress and strain components:

$$\sigma_1^e \, \varepsilon_1^e + \sigma_2^e \, \varepsilon_2^e + \sigma_3^e \, \varepsilon_3^e = \sigma_1^N \, \varepsilon_1^N + \sigma_2^N \, \varepsilon_2^N + \sigma_3^N \, \varepsilon_3^N \qquad [2.23]$$

According to equation [2.22], the total strain energy density in the notch tip of an elasto-plastic body is the same as that in a geometrically identical elastic body subjected to the same external load.

Analogously, the ESED method can be generalized for an arbitrary multiaxial stress state as:

$$\int_0^{\varepsilon_{ij}^e} \sigma_{ij}^e d\varepsilon_{ij}^e = \int_0^{\varepsilon_{ij}^E} \sigma_{ij}^E \, d\varepsilon_{ij}^E \qquad [2.24]$$

On the principal stress axes, equation [2.24] is written as:

$$\int_0^{\varepsilon_{ij}^e} \sigma_1^e d\varepsilon_1^e + \sigma_2^e d\varepsilon_2^e + \sigma_3^e d\varepsilon_3^e$$

$$= \int_0^{\varepsilon_{ij}^E} \sigma_1^E d\varepsilon_1^E + \sigma_2^E d\varepsilon_2^E + \sigma_3^E d\varepsilon_3^E \qquad [2.25]$$

Calculating the integral in equation [2.25] requires knowledge of the stress-strain relationship given by equation [2.21]. The final expression can be presented in a closed form by using all of the principal stress components and the equivalent stress. This is shown in reference [MRO 67].

In addition to equation [2.22] and equation [2.24], it has been observed that the contributions from each stress-strain component to the overall strain energy density at the notch tip is almost the same in both the hypothetical linear elastic body and the geometrically identical elastic-plastic body. This observation has resulted in the formulation of two additional equations that are related to the fractional contributions of the strain energy density. These equations are shown in the following sections.

2.3.2.1.3. Complete set of equations based on the generalized Neuber's rule

The basic equation for the upper bound limit, which relates the linear elastic and the elasto-plastic notch-tip strains and stresses, is the generalized Neuber rule (equation [2.22]). Since all stress components on the traction-free notch-tip surface must vanish, i.e. $\sigma_{11} = 0$, $\sigma_{12} = 0$, and $\sigma_{13} = 0$, the notch-tip stress state can always be reduced to a biaxial stress state characterized by two principal stress components (σ_2 and σ_3) associated with three principal strain components (ε_1, ε_2 and ε_3).

Thus, the whole notch-tip stress-strain problem can be reduced to the solution of a set of five nonlinear algebraic equations consisting of:

– the generalized Neuber rule:

$$\sigma_2^e \varepsilon_2^e + \sigma_3^e \varepsilon_3^e = \sigma_2^N \varepsilon_2^N + \sigma_3^N \varepsilon_3^N$$

[2.26]

– the constitutive stress-strain relationships:

$$\varepsilon_1^N = -\frac{v}{E}\left(\sigma_2^N + \sigma_3^N\right) - \frac{f(\sigma_{eq}^N)}{2\sigma_{eq}^N}\left(\sigma_2^N + \sigma_3^N\right)$$

$$\varepsilon_2^N = \frac{1}{E}\left(\sigma_2^N - v\sigma_3^N\right) + \frac{f(\sigma_{eq}^N)}{2\sigma_{eq}^N}\left(2\sigma_2^N - \sigma_3^N\right)$$

[2.27]

$$\varepsilon_3^N = \frac{1}{E}\left(\sigma_3^N - v\sigma_2^N\right) + \frac{f(\sigma_{eq}^N)}{2\sigma_{eq}^N}\left(2\sigma_3^N - \sigma_2^N\right)$$

where:

$$\sigma_{eq}^N = \sqrt{(\sigma_2^N)^2 - \sigma_2^N \, \sigma_3^N + (\sigma_3^N)^2}$$

– the fractional contribution of the total strain energy density:

$$\frac{\sigma_2^e \varepsilon_2^e}{\sigma_2^e \varepsilon_2^e + \sigma_3^e \varepsilon_3^e} = \frac{\sigma_2^N \varepsilon_2^N}{\sigma_2^N \varepsilon_2^N + \sigma_3^N \varepsilon_3^N} \qquad [2.28]$$

Equations [2.26] to [2.28] form a set of five equations from which all of the principal stress components (σ_2^N and σ_3^N) and the principal strain components (ε_1^N, ε_2^N and ε_3^N) can be calculated on the basis of data obtained from a linear elastic analysis, i.e. based on the stress components σ_2^e and σ_3^e and the strain components ε_1^e, ε_2^e and ε_3^e.

2.3.2.1.4. Complete set of equations based on the generalized ESED hypothesis

The generalized ESED relationship of equation [2.17] is the principal equation for the lower bound formulation that relates the linear-elastic and elastic-plastic notch-tip strains and stresses. Again, for the traction-free notch-tip surface the stress state can be represented by two principal stress components. The final set of equations can be written as follows:

– the generalized ESED:

$$\frac{1}{2}\left(\sigma_2^e \varepsilon_2^e + \sigma_3^e \varepsilon_3^e\right) = \frac{1}{3E}(1+v)\left(\sigma_{eq}^E\right)^2 +$$
$$+ \frac{1-2v}{6E}\left(\sigma_2^E + \sigma_3^E\right) + \int\limits_0^{\varepsilon_{eq}^{pE}} \sigma_{eq}^E d\varepsilon_{eq}^{pE} \qquad [2.29]$$

where:

$$\varepsilon_{eq}^{pE} = f\left(\sigma_{eq}^E\right)$$

and:

$$\sigma_{eq}^E = \sqrt{(\sigma_2^E)^2 - \sigma_2^E \, \sigma_3^E + (\sigma_3^E)^2}$$

– the constitutive equations:

$$\varepsilon_1^E = -\frac{\nu}{E}\left(\sigma_2^E + \sigma_3^E\right) - \frac{f(\sigma_{eq}^E)}{2\sigma_{eq}^E}\left(\sigma_2^E + \sigma_3^E\right)$$

$$\varepsilon_2^E = \frac{1}{E}\left(\sigma_2^E - \nu\sigma_3^E\right) + \frac{f(\sigma_{eq}^E)}{2\sigma_{eq}^E}\left(2\sigma_2^E - \sigma_3^E\right) \qquad [2.30]$$

$$\varepsilon_3^E = \frac{1}{E}\left(\sigma_3^E - \nu\sigma_2^E\right) + \frac{f(\sigma_{eq}^E)}{2\sigma_{eq}^E}\left(2\sigma_3^E - \sigma_2^E\right)$$

– the fractional contribution of the total strain energy density:

$$\frac{\sigma_2^e \varepsilon_2^e}{\sigma_2^e \varepsilon_2^e + \sigma_3^e \varepsilon_3^e} = \frac{\sigma_2^E \varepsilon_2^E}{\sigma_2^E \varepsilon_2^E + \sigma_3^E \varepsilon_3^E} \qquad [2.31]$$

Equations [2.29] to [2.31] make it possible to determine all of the stress and strain components in the notch tip provided that the hypothetical elastic notch-tip stresses, σ_2^e and σ_3^e, and the material stress-strain curve are known.

2.3.2.2. Non-proportional loading

In the case of plastic deformation caused by non-proportional loading, the final stress-strain state is dependent on the loading path. As a result, relationships that define the local stress-strain state in a notched body subjected to multiaxial loading must be developed in an incremental form. Three stress increments and four strain increments have to be found at the notch tip for each load increment. As the principal axes of the notch-tip stress tensor rotate with respect to the notch frame of reference, all of the equations have to be written in the fixed, not principal axes.

The incremental constitutive relationship forms four independent equations. The remaining three equations that are necessary for a complete formulation of the notch-tip problem can be determined by using the strain energy criteria that were discussed above in section 2.3.2.1.3 and 2.3.2.1.4, but formulated in incremental form.

2.3.2.2.1. Material constitutive model

The most frequently used material constitutive model for incremental plasticity is the Prandtl-Reuss relationship. For an isotropic body, the Prandtl-Reuss relationship can be expressed as:

$$\Delta\varepsilon_{ij} = \frac{1+v}{E}\Delta\sigma_{ij} - \frac{v}{E}\Delta\sigma_{kk}\delta_{ij} + \frac{3}{2}\frac{\Delta\varepsilon_{eq}^{p}}{\sigma_{eq}}S_{ij} \qquad\qquad [2.32]$$

The multiaxial incremental stress-strain relation is obtained from the uniaxial stress-strain curve of equation [2.15] by relating the equivalent plastic strain to the equivalent stress, such that:

$$\Delta\varepsilon_{eq}^{P} = \frac{df(\sigma_{eq})}{d\sigma_{eq}}\Delta\sigma_{eq} \qquad\qquad [2.33]$$

In the case of the Ramberg-Osgood stress-strain relationship (equation [2.20a]) the plastic strain increment of equation [2.33] can be written as:

$$\Delta\varepsilon_{eq}^{P} = \frac{d\left(\dfrac{\sigma_{eq}}{K}\right)^{\frac{1}{n}}}{d\sigma_{eq}}\Delta\sigma_{eq} \qquad\qquad [2.33a]$$

2.3.2.2.2. Incremental Neuber's rule

It is proposed that for a given increment of external load, the corresponding increment of the total strain energy density at the notch tip in an elastic-plastic body can be approximated by that which would be obtained if the body was to hypothetically remain elastic throughout the loading history.

The total strain energy is defined as the sum of the strain energy density (as described above) and the complementary strain energy density. Mathematically, this can be written as:

$$\Delta\Omega^{e} = \Delta\Omega^{N}$$

or $\qquad\qquad\qquad\qquad\qquad\qquad$ [2.34]

$$\sigma_{ij}^{e}\Delta\varepsilon_{ij}^{e} + \varepsilon_{ij}^{e}\Delta\sigma_{ij}^{e} = \sigma_{ij}^{N}\Delta\varepsilon_{ij}^{N} + \varepsilon_{ij}^{N}\Delta\sigma_{ij}^{N}$$

Equation [2.34] is called as the incremental Neuber relation since it reduces to Neuber's rule in its original form of equation [2.14] for uniaxial notch-tip stress states. Furthermore, it represents a statement of equality between the increment of the total notch-tip strain energy density that is obtained from a linear-elastic solution and the corresponding increment that is obtained from an elastic-plastic analysis.

A graphical representation of the incremental Neuber's rule is shown in Figure 2.14a, where the horizontal and vertical rectangles represent the strain energy increment for an associated pair of stress and strain components.

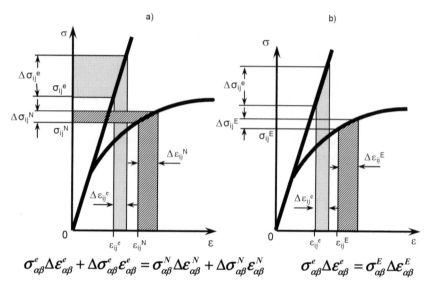

$$\sigma^e_{\alpha\beta}\Delta\varepsilon^e_{\alpha\beta} + \Delta\sigma^e_{\alpha\beta}\varepsilon^e_{\alpha\beta} = \sigma^N_{\alpha\beta}\Delta\varepsilon^N_{\alpha\beta} + \Delta\sigma^N_{\alpha\beta}\varepsilon^N_{\alpha\beta} \qquad \sigma^e_{\alpha\beta}\Delta\varepsilon^e_{\alpha\beta} = \sigma^E_{\alpha\beta}\Delta\varepsilon^E_{\alpha\beta}$$

Figure 2.14. *Graphical representation: a) incremental Neuber's rule; and b) incremental ESED method*

The four constitutive relationships used in conjunction with the generalized Neuber's rule (equation [2.34]) will only be sufficient to describe five of the seven unknown notch-tip stress-strain increments. Therefore, two more independent equations are required to completely define the notch-tip stress and strain increments for a given increment of the applied load.

Again, it is has been proposed [SIN 96] that the contribution of each elastic-plastic stress-strain component to the total strain energy density at the notch tip is equal to the contribution of the same increment of the stress-strain component when it is assumed that the body remains elastic during the loading history. This proposal can be expressed as:

$$\sigma^e_{\alpha\beta}\Delta\varepsilon^e_{\alpha\beta} + \varepsilon^e_{\alpha\beta}\Delta\sigma^e_{\alpha\beta} = \sigma^N_{\alpha\beta}\Delta\varepsilon^N_{\alpha\beta} + \varepsilon^N_{\alpha\beta}\Delta\sigma^N_{\alpha\beta} \qquad [2.35]$$

where α and β = 1, 2, 3, but summation does not apply.

Equation [2.35] in general represents six independent equations that result in equation [2.34] when they are added together side-by-side. The three equations implied by equation [2.35] and the four constitutive equations of equation [2.32] form a set of seven independent equations that are sufficient to determine the unknown increments $\Delta\sigma_{ij}{}^N$ and $\Delta\varepsilon_{ij}{}^N$ in the notch tip.

– Incremental constitutive equations:

$$\Delta\varepsilon_{11}^N = -\frac{v}{E}(\Delta\sigma_{22}^N + \Delta\sigma_{33}^N) - \frac{1}{2}(\sigma_{22}^N + \sigma_{33}^N)\frac{\Delta\varepsilon_{eq}^{pN}}{\sigma_{eq}^N}$$

$$\Delta\varepsilon_{22}^N = \frac{1}{E}(\Delta\sigma_{22}^N - v\Delta\sigma_{33}^N) + \frac{1}{2}(2\sigma_{22}^N - \sigma_{33}^N)\frac{\Delta\varepsilon_{eq}^{pN}}{\sigma_{eq}^N}$$

$$\Delta\varepsilon_{22}^N = \frac{1}{E}(\Delta\sigma_{33}^N - v\Delta\sigma_{22}^N) + \frac{1}{2}(2\sigma_{33}^N - \sigma_{22}^N)\frac{\Delta\varepsilon_{eq}^{pN}}{\sigma_{eq}^N} \qquad [2.36]$$

$$\Delta\varepsilon_{23}^N = \frac{1+v}{E}\Delta\sigma_{23}^N + \frac{3}{2}\frac{\Delta\varepsilon_{eq}^{pN}}{\sigma_{eq}^N}\sigma_{23}^N$$

where:

$$\left(\sigma_{eq}^N\right)^2 = \left(\sigma_{22}^N\right)^2 + \left(\sigma_{33}^N\right)^2 - \sigma_{22}^N\sigma_{33}^N + 3\left(\sigma_{23}^N\right)^2$$

– Strain energy density equations:

$$\sigma_{22}^e\Delta\varepsilon_{22}^e + \varepsilon_{22}^e\Delta\sigma_{22}^e = \sigma_{22}^N\Delta\varepsilon_{22}^N + \varepsilon_{22}^N\Delta\sigma_{22}^N$$

$$\sigma_{33}^e\Delta\varepsilon_{33}^e + \varepsilon_{33}^e\Delta\sigma_{33}^e = \sigma_{33}^N\Delta\varepsilon_{33}^N + \varepsilon_{33}^N\Delta\sigma_{33}^N \qquad [2.37]$$

$$\sigma_{23}^e\Delta\varepsilon_{23}^e + \varepsilon_{23}^e\Delta\sigma_{23}^e = \sigma_{23}^N\Delta\varepsilon_{23}^N + \varepsilon_{23}^N\Delta\sigma_{23}^N$$

2.3.2.2.3. Incremental ESED relations

It has been proposed [SIN 96] that for a given increment of external load, the corresponding increment in the strain energy density at the notch tip in an elastic-plastic body can be approximated by that which would be obtained if the body was to remain elastic throughout the loading history. This hypothesis can be expressed as:

$$\Delta W^e = \Delta W^E$$

or $\qquad\qquad\qquad\qquad\qquad\qquad\qquad\qquad\qquad\qquad$ [2.38]

$$\sigma_{ij}^e\Delta\varepsilon_{ij}^e = \sigma_{ij}^E\Delta\varepsilon_{ij}^E$$

Equation [2.38] represents a statement of equality between the increment of notch-tip strain energy density obtained from a linear-elastic solution and that obtained from an elastic-plastic analysis. A graphical representation of the incremental ESED method is shown in Figure 2.14b, where the strain energy densities are represented by the vertical bars of the trapezoidal shape.

The four constitutive relations used in conjunction with the generalized ESED equation will only be sufficient to formulate a set of five equations. Therefore, two more independent equations are required to completely define the notch-tip stress-strain increments for a given increment in the applied load.

Based on observations of numerical and experimental data, the following hypothesis was proposed, which states that the equality of energies also applies to all corresponding stress and strain components:

$$\sigma_{\alpha\beta}^{e}\Delta\varepsilon_{\alpha\beta}^{e} = \sigma_{\alpha\beta}^{E}\Delta\varepsilon_{\alpha\beta}^{E} \qquad [2.39]$$

Note that in equation [2.39], the indices α, β = 1, 2 and 3, but summation is not implied. The three non-zero relations given by this equation, and the four constitutive equations, are sufficient to determine the three unknown stress and the four unknown strain increments.

The complete set of incremental equations based on the equality of strain energy increments (ESED) is given below:

– Incremental constitutive relations:

$$\Delta\varepsilon_{11}^{E} = -\frac{v}{E}(\Delta\sigma_{22}^{E} + \Delta\sigma_{33}^{E}) - \frac{1}{2}(\sigma_{22}^{E} + \sigma_{33}^{E})\frac{\Delta\varepsilon_{eq}^{pE}}{\sigma_{eq}^{E}}$$

$$\Delta\varepsilon_{22}^{E} = \frac{1}{E}(\Delta\sigma_{22}^{E} - v\Delta\sigma_{33}^{E}) + \frac{1}{2}(2\sigma_{22}^{E} - \sigma_{33}^{E})\frac{\Delta\varepsilon_{eq}^{pE}}{\sigma_{eq}^{E}}$$

$$\Delta\varepsilon_{33}^{E} = \frac{1}{E}(\Delta\sigma_{33}^{E} - v\Delta\sigma_{22}^{E}) + \frac{1}{2}(2\sigma_{33}^{E} - \sigma_{22}^{E})\frac{\Delta\varepsilon_{eq}^{pE}}{\sigma_{eq}^{E}} \qquad [2.40]$$

$$\Delta\varepsilon_{23}^{E} = \frac{1+v}{E}\Delta\sigma_{23}^{E} + \frac{3}{2}\frac{\Delta\varepsilon_{eq}^{pE}}{\sigma_{eq}^{E}}\sigma_{23}^{E}$$

where:

$$\left(\sigma_{eq}^{E}\right)^{2}=\left(\sigma_{22}^{E}\right)^{2}+\left(\sigma_{33}^{E}\right)^{2}-\sigma_{22}^{E}\sigma_{33}^{E}+3\left(\sigma_{23}^{E}\right)^{2}$$

– Energy equations:

$$\sigma_{22}^{e}\Delta\varepsilon_{22}^{e}=\sigma_{22}^{E}\Delta\varepsilon_{22}^{E}$$
$$\sigma_{33}^{e}\Delta\varepsilon_{33}^{e}=\sigma_{33}^{E}\Delta\varepsilon_{33}^{E}$$
$$\sigma_{23}^{e}\Delta\varepsilon_{23}^{e}=\sigma_{23}^{E}\Delta\varepsilon_{23}^{E}$$

[2.41]

In order to determine the notch-tip elastic-plastic strains and stresses at the end of the loading history, they must first be evaluated for each increment of the applied load. Initially, the first reference state is taken as the point at which yielding occurs at the notch tip, since it can be found from an elastic analysis of the body.

For each increment of external load, the increments in the elastic-plastic notch-tip strains and stresses are computed from either equations [2.36] and [2.37], or equations [2.40] and [2.41] along with the hypothetical elastic notch-tip stress-strain history and the material stress-strain curve. The stress and strain states at the end of a given load increment are then computed using:

$$\sigma_{ij}^{n}=\sigma_{ij}^{o}+\sum_{k=1}^{n-1}\Delta\sigma_{ij}+\Delta\sigma_{ij}^{n}$$

[2.42]

$$\varepsilon_{ij}^{n}=\varepsilon_{ij}^{o}+\sum_{k=1}^{n-1}\Delta\varepsilon_{ij}+\Delta\varepsilon_{ij}^{n}$$

[2.43]

where n denotes the number of the load increment.

2.3.2.2.4. Incremental cyclic plasticity model

The set of equations defined by the incremental Neuber's rule and ESED method can only be solved if the relationship between the equivalent plastic strain increment, $\Delta\varepsilon_{eq}^{p}$, and the equivalent stress increment, $\Delta\sigma_{eq}$, is known during the application of a given load increment. It is known that the current $\Delta\varepsilon_{eq}^{p}-\Delta\sigma_{eq}$ relationship depends on the previous load path. Therefore, the use of the incremental Neuber's rule or the incremental ESED method must be associated with a plasticity model that deals with path-dependent material behavior.

Several models are available in the literature. The model proposed by Mroz [MRO 67] and more recently improved by Garud [GAR 81] is the most popular. Mroz [MRO 67] proposed that the uniaxial stress-strain material curve can be represented by a set of work-hardening surfaces in 3D stress space (Figure 2.15).

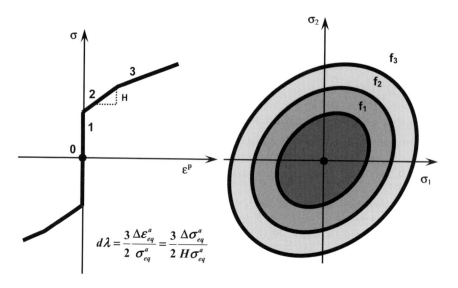

Figure 2.15. *Piecewise linearization of the material $\sigma - \varepsilon$ curve and the corresponding work-hardening surfaces*

In the case of a 2D stress state, such as that of a notch tip, the work-hardening surfaces are represented by ellipses on the coordinate plane for which the axes are defined by the principal stress components. The equation of each work-hardening surface is defined by:

$$\sigma_{eq} = \sqrt{\left(\sigma_2\right)^2 - \sigma_2\sigma_3 + \left(\sigma_3\right)^2}$$

[2.44]

The load path dependency effects are modeled by prescribing a translation rule for the motion of the ellipses that move with respect to each other over distances given by the stress increments. The ellipses move within the boundaries of each other, but they do not intersect. If one ellipse comes into contact with another, they move together as one rigid body.

The translation rule proposed by Garud [GAR 81] avoids the intersection of the ellipses that can occur with the original Mroz [MRO 67] model when the load path is defined by finite stress increments (as in the case of a numerical solution). The Garud translation rule is illustrated in Figure 2.16 and can be described by considering only two work-hardening surfaces (two ellipses) as follows:

– the line of action of the stress increment, $\Delta\sigma$, is extended to intersect the next larger non-active surface, f_2, at point B_2;

– point B_2 is connected to the center, O_2, of the surface f_2;

– a line is extended through the center of the smaller active surface, O_1, parallel to the line O_2B_2 to find point B_1 on surface f_1;

– the conjugate points B_1 and B_2 are connected by the line B_1B_2;

– surface f_1 is translated from point O_1 to point O_1', such that vector O_1O_1' is parallel to line B_1B_2. The translation is complete when the end of the vector defined by the stress increment, $\Delta\sigma$, lies on the translated surface f_1'.

The translation rule described above assures that the two ellipses will be tangent to each other at the common point B_1B_2 without intersecting. Two or more tangent ellipses translate together as rigid bodies and the largest moving ellipse (see Figure 2.16) indicates which linear element of the constitutive stress-strain curve should be used for a given stress-strain increment.

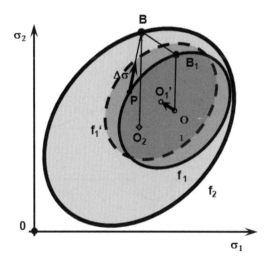

Figure 2.16. *Geometrical illustration of the Mroz-Garud incremental plasticity model [GAR 81, MRO 67]*

In most publications, the plasticity models are described as algorithms for calculating strain increments that result from given stress increments or vice versa.

In the case of the notch analysis described above, both the strain and stress increments are determined from either the Neuber or the ESED method. The plasticity model is therefore only needed to indicate which work-hardening surface is active during the next load increment.

The Mroz-Garud model [MRO 67] was chosen here as an illustration. Obviously, any other plasticity model can be associated with the incremental Neuber or ESED method.

2.4. Comparison of notch stress-strain calculations with numerical data

2.4.1. *Monotonic proportional loading*

The accuracy of the proposed method is demonstrated by comparison with the finite element data of Seeger and Hoffman [SEE 86]. These authors analyzed a cylindrical specimen (see Figure 2.17) with a circumferential notch subjected to simultaneous proportional tensile and torsion loading.

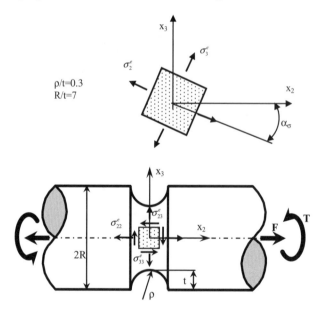

Figure 2.17. *Geometry and dimensions of the cylindrical specimen analyzed by Seeger and Hoffmann [SEE 86]*

The load in the finite element calculations was applied in such a way that the ratio of the nominal notch-tip shear to tensile stress was $\tau_n/\sigma_{nF} = 0.411$. The nominal stresses were determined from the net cross-section dimensions:

$$\sigma_{n.s.} = \frac{P}{\pi(R-t)^2} \; ; \; \tau_{n.s.} = \frac{2T}{\pi(R-t)^3} \tag{2.45}$$

The stress concentration factors for tension and torsion were $K_F = 3.89$ and $K_T = 2.19$ respectively. The stress concentration factors were defined (see Figure 2.17) as:

$$K_F = \frac{\sigma_{22}^e}{\sigma_{nF}} \quad \text{and} \quad K_T = \frac{\sigma_{32}^e}{\tau_n} \tag{2.46}$$

The ratio of the notch-tip hoop stress to the axial stress in tension was $\sigma_{33}^e/\sigma_{22}^e = 0.2$. A bi-linear stress-strain relation was used for the calculations. Its mathematical form was given by:

$$\varepsilon = \frac{\sigma}{E} \quad \text{for} \quad \sigma \leq \sigma_0 \tag{2.47}$$

and:

$$\varepsilon = \frac{\sigma_o}{E} + \frac{\sigma - \sigma_o}{H} \quad \text{for} \quad \sigma > \sigma_o \tag{2.48}$$

where:

- $E = 94{,}400$ MPa;

- $H = (0.005)E$;

- $\sigma_o = 550$ MPa; and

- $v = 0.3$.

Both the Neuber (equations [2.26] to [2.28]) and the ESED (equations [2.29] to [2.31]) method were used for calculating the notch-tip strains, ε_{ij}^E and ε_{ij}^N, and stresses, σ_{ij}^E and σ_{ij}^N. The solution method for both sets of equations is given in reference [REI 97].

The calculated strains and stresses were compared (see Figures 2.18 and 2.19) with the Seeger-Hoffman [REI 97] elasto-plastic finite element data, where the nominal equivalent stress, S, was used as the reference:

$$S = \sqrt{\sigma_{nF}^2 + 3\tau_n^2} \qquad\qquad [2.49]$$

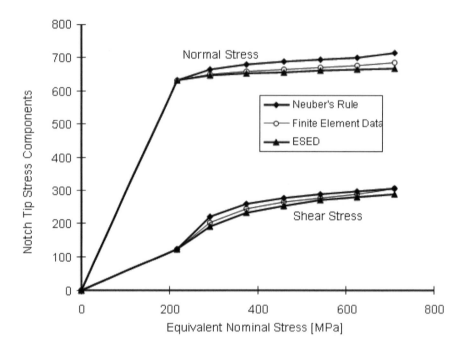

Figure 2.18. *Comparison of the Neuber and ESED stress solutions with Seeger and Hoffmann's multiaxial elastic-plastic finite element data [SEE 86]*

It has been found that the finite element data were within or close to the band predicted by the Neuber and ESED methods. It is believed that these methods give the upper and lower bounds for the elasto-plastic strains and stresses in the notch tip. However, due to numerical errors, the actual stresses might sometimes be close to but outside the band. This is especially true for the principal stress component with the smallest magnitude.

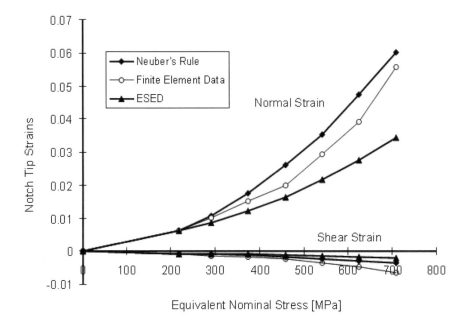

Figure 2.19. *Comparison of the Neuber and ESED strain solutions with Seeger and Hoffmann's multiaxial elastic-plastic finite element data [SEE 86]*

2.4.2. *Monotonic non-proportional loading*

The accuracy of the proposed incremental Neuber's rule and incremental ESED method were assessed by comparing the calculated notch-tip stress-strain histories to those obtained using the finite element method. The elastic-plastic results from the finite element analysis of [SIN 96] were obtained using the Abaqus [REI 97].

The geometry of the notched element was that of a circumferentially notched bar similar to that shown in Figure 2.17. The loads applied to the bar were torsion and tension according to the path shown in Figure 2.20. The nominal torsion stresses, τ_n, and tensile stresses, σ_{nF}, were determined based on the net cross-section according to equation [2.45]. The basic proportions of the cylindrical component were $\rho/t = 0.3$ and $R/t = 7$, resulting in the torsion and tensile stress concentration factor $K_T = 3.31$ and $K_F = 1.94$, respectively. The ratio of the notch-tip hoop to axial stress under tensile loading was $\sigma_{33}^e/\sigma_{22}^e = 0.284$.

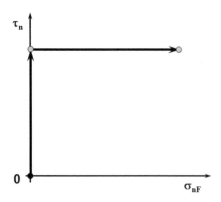

Figure 2.20. *The non-proportional monotonic torsion-tension load path*

The material for the notched bar was SAE 1045 steel with a cyclic stress-strain curve approximated by the Ramberg-Osgood relation:

$$\varepsilon = \frac{\sigma}{E} + \left(\frac{\sigma}{k'}\right)^{\frac{1}{n'}}$$
[2.50]

The material properties were:

– $E = 202,000$ MPa;

– $v = 0.3$;

– $S_y = 202$ MPa;

– $n' = 0.208$; and

– $K' = 1,258$ MPa.

The maximum applied load levels were chosen to be 50% higher than that required to cause yielding at the notch tip if each load were applied separately.

Specifically, the maxima were $\sigma_{nF}{}^f = 103$ MPa and $\tau_n{}^f = 90$ MPa. The final ratio of the nominal stresses was $\sigma_{nF}{}^f/\tau_n{}^f = 1.133$. The normalized nominal equivalent net sectional stress ratio was defined as:

$$\frac{S_{eq}}{S_y} = \frac{\sqrt{\sigma_{n.s.}^2 + 3\tau_{n.s}^2}}{S_y}$$
[2.51]

and reached a value of 0.92 at the end loading path, a value that indicates almost general yielding of the net section.

The history of the linear elastic stress at the notch tip that was induced by the applied remote load path is shown in Figure 2.20.

The history of the normal and shear stress was used together with equations [2.36], [2.37], [2.40] and [2.41] in order to calculate the notch-tip elastic-plastic strains and stresses. The calculated strains and stresses were subsequently compared to the elastic-plastic finite element data. The strain components (ε_{22} and $\varepsilon_{23}\varepsilon_{22}$) and the stress components (σ_{22} and σ_{23}) that were calculated according to the method described above are shown in Figures 2.21 and 2.22.

Note that results from both approximate methods and the finite element analysis are identical in the elastic range. This is expected since the models converge to the elastic solution in the elastic range.

Just beyond the onset of yielding at the notch tip, the strain results that were predicted using the proposed models and the finite element data begin to diverge.

It can be concluded that the incremental Neuber's method predicts an upper bound, and the incremental ESED method a lower bound approximation to the actual notch-tip strains.

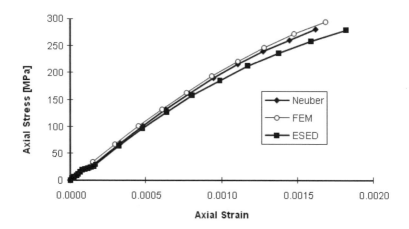

Figure 2.21. *Comparison of the Neuber and ESED normal stress-strain solutions with the multiaxial elastic-plastic finite element method (FEM) data of Seeger and Hoffmann [SEE 86]*

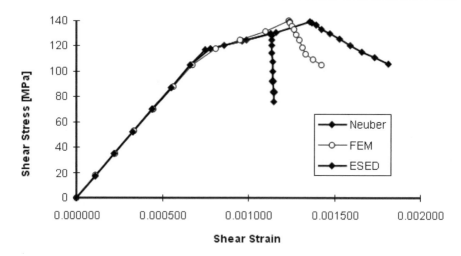

Figure 2.22. *Comparison of the Neuber and ESED shear stress-strain solutions with the multiaxial elastic-plastic finite element method (FEM) data of Seeger and Hoffmann [SEE 86]*

2.4.3. *Proportional multiaxial cyclic loading*

The accuracy of the proposed methods in the case of proportional multiaxial cyclic loading is demonstrated by comparison with the finite element data of Hoffman [SEE 86], who analyzed the cylindrical specimen shown in Figure 2.17.

The load in the finite element calculations was applied in such a way that the ratio of the nominal notch-tip shear to tensile stress was $\tau_n / \sigma_{nF} = 2.5$. The nominal stresses were determined from the net cross-section dimensions according to equation [2.45] and the stress concentration factors $K_F = 3.89$ and $K_T = 2.19$ were defined according to equation [2.46].

The ratio of notch-tip hoop stress to axial stress in tension was $\sigma_{33}^e / \sigma_{22}^e = 0.2$.

A cyclic loading history (see Figure 2.23), consisting of eight reversal points, was used for the analysis (see Figure 2.14). The load values at the reversal points were defined as the ratio of the equivalent nominal stress, S (equation [2.49]) to the yield stress, σ_0, of the material used for the analysis, (i.e. $S/\sigma_0 = 0.0$, 0.8, -0.2, 1.0, -1.0, 0.6, -0.2 and 1.0). The entire load history is shown in Figure 2.23. The same material stress-strain curve (equations [2.47] and [2.48]) was used as in the case of the analysis previously carried out for the proportional loading path.

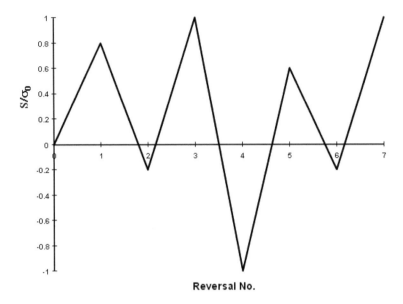

Figure 2.23. *The tension-torsion cyclic loading history of*
Seeger and Hofmann [SEE 86]

Both the Neuber (equations [2.26] to [2.28]) and ESED (equations [2.29] to [2.31]) method were used for calculating the notch-tip strains, ε_{ij}^E and ε_{ij}^N, and stresses, σ_{ij}^E and σ_{ij}^N. The calculations were carried out for each hysteresis loop, $\sigma - \varepsilon$, identified by the *rainflow* counting procedure [GLI 87].

Each branch of the stress-strain hysteresis loop was described by the original stress-strain curve (equations [2.47] and [2.48]) expanded by a factor of two, as is usually done in the case of uniaxial stress state fatigue analysis [WET 77]. The calculated principal strains and stresses were compared (Figures 2.24 and 2.25) with the Hoffman results [SEE 86]. Only the leading stress component, σ_2, and the corresponding strain, ε_2, are shown in Figures 2.24 and 2.25 respectively. The stress, σ_2, was less than the uniaxial yield limit but plastic yielding did take place at the notch tip due to the joint action of all stress components, i.e. σ_2 and σ_3.

The comparison of the calculated notch-tip strains and stresses revealed that the Neuber-type relationships (equations [2.26] to [2.28]) over-predicted the elastic-plastic notch-tip strains and stresses while the ESDE equations (equations [2.29] to [2.31]) resulted in under prediction. This tendency was found to be consistent over a variety of geometry and loading configurations. The dominant strain and stress components were most often found within the limits defined by the Neuber-type and ESED relationships.

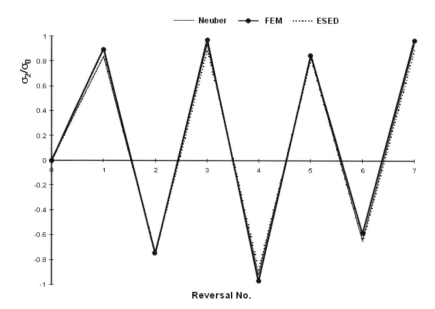

Figure 2.24. *Comparison of the dominant principal stress component σ_2 calculated at the notch tip with Hoffman's finite element method (FEM) data [SEE 86]*

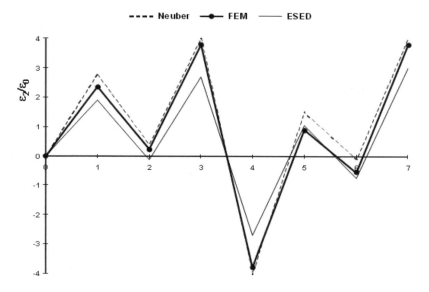

Figure 2.25. *Comparison of the calculated dominant principal strain component ε_2 at the notch tip with Hoffman's finite element data [SEE 86]*

The elastic-plastic strains and stresses at the notch tip can subsequently be used for calculating the fatigue damage by inputting them into appropriate multiaxial fatigue damage parameter.

2.5. Conclusion

Two methods for calculating elasto-plastic notch-tip strains and stresses have been proposed for general multiaxial loading. The methods have been formulated using both the total strain energy density and the strain energy density relationships in conjunction with the Mroz-Garud cyclic plasticity model. It has been found that the generalized Neuber's rule, which represents the equality of the total strain energy density at the notch tip, gives an upper bound estimate for the elasto-plastic notch-tip strains. The generalized equation of the equivalent strain energy density yields a lower-bound solution for the notch-tip strains and stresses. The method has been verified by comparison to numerical and experimental data for various loading paths and material properties.

The calculated notch-tip strains and stresses were subsequently used for calculating fatigue damage according to the proposed multiaxial fatigue damage parameter that associates strain energy density with the critical plane concept.

2.6. Bibliography

[BAN 92] J.A. BANNATINE, D.F. SOCIE, "Multiaxial fatigue life estimation technique", in: M.R. MITCHELL and R.W. LANDGRAF (eds), *Symposium on Multiaxial Fatigue*, ASTM STP 1122, Philadelphia, pp. 249-275, 1992.

[BAR 94] M.E. BARKEY, D.F. SOCIE, K.J. HSIA, "A yield surface approach to the estimation of notch strains for proportional and non-proportional cyclic loading", *ASME Journal of Engineering Materials and Technology*, vol. 116, pp. 173-180, 1994.

[CHU 92] C.C. CHU, F. CONLE, J.J. BONNEN, "Multiaxial stress-strain modelling and fatigue life prediction of SAE axle shafts", in: M.R. MITCHELL and R.W. LANDGRAF (eds), *Symposium on Multiaxial Fatigue*, ASTM STP 1122, Philadelphia, pp. 249-275, 1992.

[DAN 89] K. DANG VAN, B. GRIVEAU, O. MESSAGE, "On a new mutiaxial fatigue limit criterion: theory and application", *Biaxial and Multiaxial Fatigue*, EFG Publication 3, Mechanical Engineering Publications, London, pp. 479-496, 1989.

[ELL 84] F. ELLYIN, D. KUJAWSKI, "Plastic strain energy in fatigue failure", *Journal of Engineering Materials and Technology*, vol. 106, pp. 342-347, 1984.

[ELL 88] F. ELLYIN, K. GOLOS, "Multiaxial fatigue damage criterion", *Journal of Engineering Materials and Technology*, vol. 110, pp. 63-68, January 1988.

[FAS 85] J.W. FASH, An Evaluation of Damage Development During Multiaxial Fatigue of Smooth and Notched Specimens, Report No. 123/UILU-ENG 85-3607, University of Illinois at Urbana-Champaign, 1985.

[FAT 88] A. FATEMI, P. KURATH, "Multiaxial fatigue life predictions under the influence of mean-stress", *Journal of Engineering Materials and Technology*, vol. 110, pp. 380-388, 1988.

[FEL 61] C.E. FELTNER, J. MORROW, "Microplastic strain hysteresis energy as a criterion for fatigue fracture", *Journal of Basic Engineering,* vol. 83D, pp. 15-22, 1961.

[FIN 53] W.N. FINDLEY, Combined Stress Fatigue Strength of 76S-T61 with Super-imposed Mean Stresses and Corrections for Yielding, Technical Note 2924, NACA-National Advisory Committee for Aeronautics, 1953.

[FIN 59] W.N. FINDLEY, "Theory for the effect of mean stress on fatigue of metals under combined torsion and axial load or bending", *Journal of Engineering for Industry*, vol. 81, pp. 301-306, 1959.

[FIN 61] W.N. FINDLEY, P.N. MATHUR, E. SZCZEPANSKI, A.O. TEMEL, "Energy versus stress theories for combined stress - A fatigue experiment using a rotating disk", *Journal of Basic Engineering*, vol. 83D, pp. 10-14, 1961.

[GAR 79] Y.S. GARUD, "A new approach to the evaluation of fatigue under multi-axial loadings", *Methods for Predicting Material Life in Fatigue*, ASME, New York, pp. 247-263, 1979.

[GAR 81] Y.S. GARUD, "A new approach to the evaluation of fatigue under multiaxial loading", *Journal of Engineering Materials and Technology*, vol. 103, pp. 118-125, 1981.

[GLI 87] G. GLINKA, J.C.P. KAM, "Rainflow counting algorithm for very long stress histories", *International Journal of Fatigue*, vol. 9, no. 3, pp. 223-228, 1987.

[GLI 95a] G. GLINKA, A. PLUMTREE, G. SHEN, "A multiaxial fatigue strain energy density parameter related to the critical plane", *Journal of Fatigue and Fracture of Engineering Materials and Structures*, vol. 18, pp. 37-46, 1995.

[GLI 95b] G. GLINKA, G. WANG, A. PLUMTREE, "Mean stress effects in multiaxial fatigue", *Journal of Fatigue and Fracture of Engineering Materials and Structures*, vol. 18, pp. 755-764, 1995.

[JOR 85] E.H. JORDAN, M.W. BROWN, K.J. MILLER, "Fatigue under severe nonproportional loading", in K.J. MILLER and M.W. BROWN (eds), *Multiaxial Fatigue*, ASTM STP 853, Philadelphia, pp. 569-585, 1985.

[KAL 02] A.R. KALLMEYER, A. KRGO, P. KURATH, "Evaluation of multiaxial fatigue life prediction methodologies for Ti-6Al-4V", ASME *Journal of Engineering Materials and Technology*, vol. 124, no. 2, pp. 229-237, 2002.

[KAN 82] F.A. KANDIL, M.W. BROWN, K.J. MILLER, "Biaxial low cycle fatigue of 316 stainless steel at elevated temperature", *The Metals Society*, London, vol. 280, pp. 203-210, 1982.

[KOC 85] J.L. KOCH, Proportional and Non-proportional Biaxial Fatigue of Inconel 718, Report No. 121/UILU-ENG, 85-3605, University of Illinois at Urbana-Champaign, 1985.

[KUR 88] P. KURATH, S. DOWNING, D.R. GALLIART, "Summary of non-hardened notched shaft round robin program", in: G.E. LEESE and D. SOCIE (eds), *Multiaxial Fatigue, Analysis and Experiments*, Society of Automotive Engineers, Warrendale, PA, pp. 13-31, 1988.

[LOH 80] R.D. LOHR, E.G. ELLISON, "A simple theory for low cycle multiaxial fatigue", *Journal of Fatigue and Fracture of Engineering Materials and Structures*, vol. 3, pp. 1-17, 1980.

[MCD 91] D.L. MC DIARMID, "A general criterion of high cycle multiaxial fatigue failure", *Journal of Fatigue and Fracture of Engineering Materials and Structures*, vol. 14, pp. 429-453, 1991.

[MOF 95] A. MOFTAKHAR, A. BUCZYNSKI, G. GLINKA, "Calculation of elasto-plastic strains and stresses in notches under multiaxial loading", *International Journal of Fracture*, vol. 70, pp. 357-373, 1995.

[MOL 81] K. MOLSKI, G. GLINKA, "A method of elastic-plastic stress and strain calculation at a notch root", *Material Science and Engineering*, vol. 50, pp. 93-100, 1981.

[MRO 67] Z. MROZ, "On the description of anisotropic workhardening", *Journal of Mechanics and Physics of Solids*, vol. 15, pp. 163-175, 1967.

[NEU 61] H. NEUBER, "Theory of stress concentration shear strained prismatic bodies with arbitrary non- linear stress-strain law", ASME *Journal of Applied Mechanics*, vol. 28, pp. 544-550, 1961.

[OTT 87] H. OTT, Neues Verfahen zur Betriebfestigkeitsanalyse komplexer Bauteile auf der Grundlage der FEM, PhD Thesis, University of Aachen (RWTH), Germany, 1987.

[RAD 80] V. M. RADAKRISHNAN, "An analysis of low cycle fatigue based on hysteresis energy", *Journal of Fatigue & Fracture of Engineering Materials and Structures*, vol. 3, pp. 75-84, 1980.

[REI 97] W. REINHARDT, A. MOFTAKHAR, G. GLINKA, "An efficient method for calculating multiaxial elasto-plastic notch tip strains and stresses under proportional loading", *27th ASTM Fracture Mechanics Conference, ASTM STP vol 1296*, Blacksburg, Virginia, pp. 613-629, 1997.

[SEE 86] T. SEEGER, M. HOFFMAN, The Use of Hencky's Equations for the Estimation of Multiaxial Elastic-Plastic Notch Stresses and Strains, Report No. FB-3/1986, Technische Hochschule Darmstadt, 1986.

[SIN 59] G. SINES, "Behaviour of metals under complex static and alternating stresses", in: G. SINES and J.L. WAISMAN (eds), *Metal Fatigue*, McGraw-Hill, New York, pp. 145-169, 1959.

[SIN 96] M.N.K. SINGH, G. GLINKA, R.N. DUBEY, "Elastic-plastic stress-strain calculation in notched bodies subjected to non-proportional loading", *International Journal of Fracture*, vol. 76, no. 1, pp. 39-60, 1996.

[SOC 89] D.F. SOCIE, P. KURATH, J. KOCH, "A multiaxial fatigue damage parameter", in: M.W. BROWN and K.J. MILLER, *Biaxial and Multiaxial Fatigue-EGF3*, Mechanical Engineering Publications, London, pp. 535-550, 1989.

[SOC 00] D.F. SOCIE, G.B. MARQUIS, *Multiaxial Fatigue*, Warrendale, PA, 2000.

[STE 97] S. STEFANOV, "The curvilinear integral method: a question to the critical plane concept", *International Journal of Fatigue*, vol. 19, no. 2, pp. 101-107, 1997.

[STE 02] S. STEFANOV, "Fatigue life prediction without cycle counting (by means of the integral method)", *Journal of Theoretical and Applied Mechanics*, vol. 32, no. 1, pp. 34-47, 2002.

[TIP 88] S.M. TIPTON, J.W. FASH, "Multiaxial fatigue life predictions of the SAE specimen using strain based approaches", *Multiaxial Fatigue, Analysis and Experiments*, AE-14, pp. 67-80, 1988.

[TIP 89] S.M. TIPTON, D.V. NELSON, "Multiaxial fatigue life predictions for the SAE specimen using stress based approaches", in: G.E. LEESE and D. SOCIE (eds), *Multiaxial Fatigue: Analysis and Experiments*, Society of Automotive Engineers, Warrendale, PA, pp. 61-65, 1989.

2.7. Symbols

α, β	indices. $\alpha, \beta = 1,2,3$, summation is not implied
α_σ	angle of principal axis with respect to the specimen axis
$\Delta\varepsilon_{22}$	normal strain range in the critical plane
$\Delta\varepsilon_{ij}^P$	plastic strain increments
$\Delta\gamma_{21}$	shear strain range in the critical plane
δ_{ij}	Kronecker delta, $\delta_{ij} = 1$ for $i = j$ and $\delta_{ij} = 0$ for $i \neq j$
ΔP	axial load range
$\Delta\sigma_{21}$	shear stress range in the critical plane
$\Delta\sigma_{22}$	normal stress range in the critical plane
$\Delta\sigma_{ij}, \Delta\varepsilon_{ij}$	increments in notch-tip stress and strain components
ΔT	torsion moment range
ΔW^e	elastic strain energy density
ΔW^p	plastic strain energy density
E	modulus of elasticity

$\varepsilon_{22}{}^{a}$	amplitude of normal strain component in the critical plane
$\varepsilon_{eq}{}^{p}$	equivalent plastic strain
$\varepsilon_{eq}{}^{pE}$	equivalent plastic strain determined from the ESED method
$\varepsilon_{ij}{}^{a}$	actual elasto-plastic strain components in the notch tip
$\varepsilon_{ij}{}^{A}$	strain components in fixed coordinate system coaxial with the specimen axis
$\varepsilon_{ij}{}^{E}$	elasto-plastic notch-tip strains obtained from the ESED method
$\varepsilon_{ij}{}^{e}$	notch-tip strain components obtained from linear elastic analysis
$\varepsilon_{ij}{}^{N}$	elasto-plastic notch-tip strains obtained from the Neuber method
$\varepsilon_{ij}{}^{n}$	nominal strain tensor components
$\varepsilon_{ij}{}^{p}$	plastic components of the notch-tip strain tensor
ε_{n}	nominal strain
ESED	equivalent strain energy density
F	axial load
γ^{*}	multiaxial fatigue parameter
$\gamma_{21}{}^{a}$	shearing amplitude in the critical plane
H	plastic modulus
K'	cyclic strength coefficient
k, K	material constants
K_{ε}	strain concentration factor
K_{σ}	stress concentration factor
K_{t}	theoretical elastic stress concentration factor
v	Poisson's ratio
n	exponent of power law stress-strain curve
n'	cyclic strain-hardening exponent
v_{e}, v_{p}	elastic and plastic Poisson's ratios, respectively
v_{eff}	effective Poisson's ratio
N_{f}	number of cycles to failure or to the initiation of a 1 mm crack
R	radius of a cylindrical specimen
S	equivalent nominal stress
$\sigma_{21}{}^{a}$	amplitude of shear stress component in the critical plane
$\sigma_{21}{}^{max}$	maximum absolute value of shear stress in the critical plane
$\sigma_{22}{}^{a}$	amplitude of normal stress component in the critical plane
$\sigma_{22}{}^{max}$	maximum value of normal stress in the critical plane

S_{eq}	equivalent nominal stress
σ_{eq}	equivalent stress
σ_f'	axial cyclic fatigue strength coefficient
S_{ij}	deviatoric stress components
σ_{ij}	stress components associated with the critical plane
$\sigma_{ij}^{\,a}$	actual stress tensor components in the notch tip
$\sigma_{ij}^{\,A}$	stress components in fixed coordinate system coaxial with the specimen axis
$\sigma_{ij}^{\,e}$	notch-tip stress-tensor components obtained from linear-elastic analysis
$\sigma_{ij}^{\,E}$	notch-tip stress-tensor components obtained from the ESED model
$\sigma_{ij}^{\,N}$	notch-tip stress-tensor components obtained from the Neuber solution
σ_n	nominal stress
$\sigma_n^{\,F}$	nominal (average) stress in the net cross section due to axial load F
σ_o	parameter of the material stress-strain curve (yield limit)
S_y	yield stress
T	torque
t	wall thickness
τ_f'	torsional cyclic fatigue strength coefficient
τ_n	nominal shear stress in the net cross-section obtained from the torsion formula
W	strain energy density
$W*$	multiaxial fatigue strain energy density parameter
W^a	actual strain energy density at the notch tip
W^E	strain energy density at the notch tip according to the ESED hypothesis
W^e	strain energy density at the notch tip obtained from linear elastic solution
W^N	strain energy density at the notch tip according to the generalized Neuber's rule

Chapter 3

Fatigue of Composite Materials

3.1. Introduction

The term "composite materials" covers a large range of materials. Obviously, it is not always easy to find a general approach to determine the fatigue behavior of some materials, including polymer matrix, metal matrix and ceramic matrix composites, along with elastomeric composites, Glare, short fiber-reinforced polymers and nanocomposites.

For this reason, we have to set up some boundaries. In addition, it is quite useful for the engineer to carefully compare the fatigue of composite materials to the fatigue of metals, as the design of a component made of a composite material is often the result of a metal substitution. We should keep in mind what Professor Tsaï said: "Think composite".

As Bryan Harris mentioned in *Fatigue in Composites* [HAR 03]: "it is wrong to assume, *a priori*, that there are some universal mechanisms by which fluctuating loads will inevitably result in failure at stresses below the normal monotonic failure stress of the material".

Fracture within high-performance composite material structures is of a very different nature compared to that of metallic components. Failure due to fatigue is less feared in the first case than in the second.

Chapter written by Claude BATHIAS.

Metallic fatigue has been intensively studied for more than 150 years, since the pioneering work performed by Wöhler. Nowadays, metallic fatigue is studied using some well-established concepts:

– low-cycle fatigue by Coffin and Manson;

– megacycle fatigue by Wöhler;

– gigacycle fatigue by Bathias;

– crack propagation by Paris.

The fatigue of composite materials is slightly different, as gigacycle fatigue of composite materials has not yet been studied in detail, as the S-N curves are usually plotted between 10^3 and 10^7 cycles and no further [BAT 04]. We should also bear in mind that the concept of damage tolerance can only be applied to metals and not to composite materials. Delamination, which is primarily considered a defect of the plane, can be treated using the mechanics of fracture with a modified Paris' law. The application of fracture mechanics to notched laminated plates is more complex.

In some cases where a quasi-isotropic layout is observed, it is always possible to use fracture mechanics to predict the failure of notched plates. However, when a strong orthotropic effect appears, the fracture mechanics criterion does not work very well. Some other criterion, such as the Nuismer and Withney [WHI 74] one or those from damage mechanics model [LAD 03] are highly recommended. A specific modeling then has to be performed.

Figure 3.1. *First part made of carbon fibers manufactured in France for a civil aircraft in the early 1970s. Landing gear door*

Despite the lack of theoretical knowledge, composite materials now play a major role from the aerospace to mechanical engineering industry. For the last 30 years, the aerospace industry has replaced the aluminum or sometimes titanium parts, and more rarely the steel ones, with components made from composite materials (see Figure 3.1). Composites were quickly adopted in the 1980s but their use slowed down in the 1990s until Boeing's Dreamliner 787 aircraft was manufactured using composites in the 2000s.

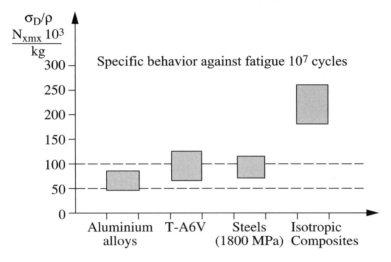

Figure 3.2. *Comparison of the specific fatigue strength of the different kinds of materials in the space industry*

High-performance composite materials, reinforced by long fibers of carbon, glass, boron or Kevlar, have a reputation for good fatigue behavior. The specific properties of these materials and their exceptional fatigue strength is responsible for their success in the aeronautic industry (see Figure 3.2). It is interesting to note, in connection with the above statement, that the substitution of aluminum alloys by carbon- and glass-fiber composites in helicopter propeller blades has improved their durability in crucial conditions.

Thirty years ago, it was common for users of carbon fiber reinforced polymer (CFRP) involved in the Airbus program to express the belief that these advanced materials did not suffer from fatigue. Today some experts, including Harris [HAR 03], assume that CFRP fatigue could be one reason, among others, to explain the crash of the AA587 flight that occurred in November 2001 near New York airport (see Figure 3.3).

Figure 3.3. *Service failure of a vertical tail at several fitting points*

In fact, the ratio S_D/UTS between the fatigue strength, S_D, in tension-tension and the ultimate tensile strength, UTS, is always >0.4 and can reach 0.9 for carbon fiber-polymer composites depending on the stacking sequence. These values can be compared to those found for metals of <0.5, and only 0.3 in the case of aluminum alloys (see Figure 3.4).

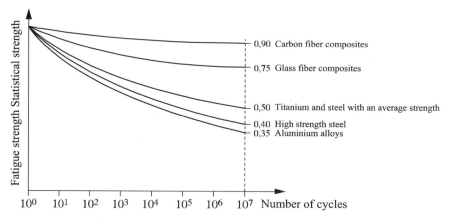

Figure 3.4. *The higher endurance of composites compared to the metals used in aeronautics*

This means that for many aeronautical components made of carbon-fiber composites, fatigue strength is not a key design parameter. For instance, the first parameter when designing a vertical tail made in carbon-epoxy is not fatigue damage. The vertical stabilizer for an aircraft is designed facing the load in standard crash conditions, which is assumed to be more severe than the conditions of service cyclic loading.

• In any case, this exceptional behavior of high-performance composite materials does not mean that they are totally sheltered from fatigue. In reality, the endurance of composites varies according to their fiber type, resin and lay up: the fatigue strength of carbon fiber is much greater than that of glass fiber; epoxy resin is better than polyester; and so on. In addition, we should mention that composites can be damaged to a much greater degree than metals in fatigue under shear or compression loadings. For instance, the simultaneous damage of a composite material under cycling bending (tension, compression and shear conditions) makes this phenomenon much more complex.

• Finally, it is well-known that the fatigue strength of composite materials is much lower in compression-compression than in tension-tension conditions. The opposite goes for metallic alloys. The most striking observation is that the fatigue strength of notched composite materials, whose ratio K_f is between the endurance limit of the untouched and notched material, is very close to 1. This means that these composite materials are not sensitive to the notch effect under cyclic stress conditions. Composite materials, which are vulnerable to the concentration of monotonous stress, remain unaffected under cyclic loading. On the other hand, compared to metals, composite materials under fatigue with a high number of cycles are very interesting when they are notched. This is because their endurance limit is two to three times higher for a given tension strength. Under low-cycle fatigue conditions, the fatigue strength of notched composites is less interesting. This is because notched composites lack the plastic accommodation of metals. It will be the same for the over-loading effects under tension conditions.

From these observations it is worth pointing out some drastic differences between metals and high-performance composites, depending on what the designer needs.

3.2. Drastic differences between the fatigue of composites and metals

Many differences occur at both the microscopic and macroscopic level to explain the fatigue behavior of high-performance composite materials. The main aspects are presented in Figure 3.5.

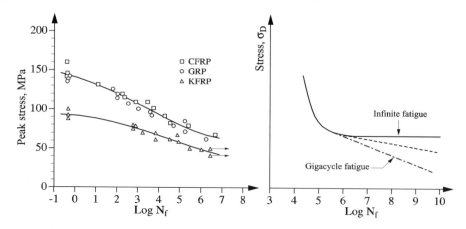

Figure 3.5. *S-N curves: composites on the left side; metals on the right side. CFRP is a carbon; GFRP a glass; KFRP Kevlar*

3.2.1. *Damage at the microscopic level*

In metals, fatigue damage is strongly related to cyclic plasticity; in other words, to the dislocation mobility and to slip systems. Due to the effect of the environment and the plane stress states, fatigue damage is often localized near the surface of the metals. Fatigue damage in composite materials is not related to plasticity (see Figures 3.6 and 3.7).

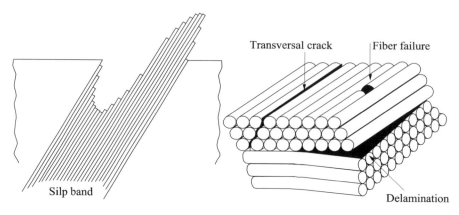

Figure 3.6. *a) Initiation models of fatigue with slip bands within metals; and b) with transversal cracks within composites*

a)   b)

Figure 3.7. *Micrographs of the initiation step showing: a) some slip bands (metals): and b) some transversal cracks (composites), magnification X750*

⸰ If we only consider polymer-matrix composites reinforced by long fibers, we usually reckon that the first damage, which occurs under loading, is cracking of the matrix before the fracture of the fibers. These cracks are microscopic, with an initial thickness of a single layer. Their appearance triggers the initiation of damage (see Figure 3.7). Propagation of the microcracks will then occur, leading to a critical crack density along with delamination, until the potential fracture of the fibers.

⸰ Compared to metals and alloys, where fatigue is localized near the surface, the microscopic mechanisms of composites are within the material itself. Every layer becomes cracked parallel to the fibers, depending on their orientation with regards to the loading.

Reifsnider [REI 87] proposed the existence of a critical crack density for a given type of layer. This hypothesis has been verified several times, but never in the case of fatigue.

For instance, Daniel [DAN 87] (see Figure 3.9) shows that within a stratified material $(0/90_2)$ made of carbon epoxides we can predict the lifetime, N_f, using the following equation:

$$N_f = (n_{CDS})^{1.25} \qquad\qquad [3.1]$$

where n_{CDS} is the number of cycles that are necessary to reach the critical crack density within the layers at 90°. This also corresponds to the end of the second damaging phase described in section 3.6.1.

This critical crack density (CDS) notion is also true within a T300/1034 $(0/\pm45/90)$ composite. In addition, the transversal crack density decreases when the humidity level increases [CAR 86]. The effect of humidity and temperature are confirmed in the study of a carbon-epoxide (0/±45) under cyclic tension-compression conditions [KEL 87].

We can therefore conclude that an increase in humidity level, which reduces both the shearing and the glass transition temperature, Tg, leads to an increase in tenacity of the matrix under tension conditions and to a decrease under compression conditions.

At the mesoscopic level, propagation of the damage under fatigue conditions within a metal is roughly a simple crack perpendicular to the loading axis. The crack tip is surrounded by a plastic zone, whose diameter about 100 microns.

Within composite materials, fatigue damage is multidirectional and the damage zone, which is much larger than the plastic zone, is related to the complex morphology of the fracture [BAT 05] (see Figure 3.8).

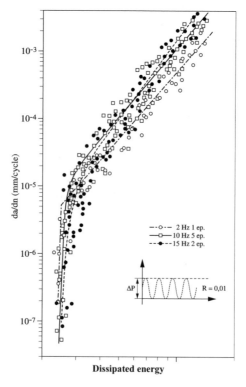

Figure 3.8a. *Delamination of composites; determination of Paris' law, dissipated energy ΔG/increase in delamination*

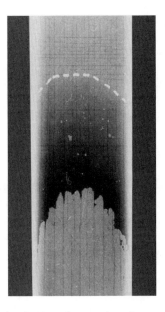

Figure 3.8b. *Delamination of composites shown using radioscopy*

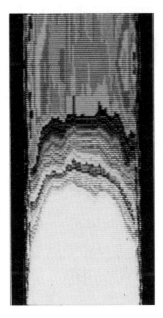

Figure 3.8c. *Delamination of composites; tomographic radiography of the delamination and the damaged zone within its core*

3.2.2. *Role of plasticity and nonlinear behavior*

The endurance curve of metallic materials, according to international normalization, is hyperbolic and significantly concave as soon as the maximum stress of the cycle reaches the elasticity limit of the material. When plasticity spreads, the lifetime (N_f) of the sample is given by the Coffin-Manson and Basquin equations:

$$\Delta\varepsilon_e/2 + \Delta\varepsilon_p/2 = a\,(2N_f)^b + b\,(2N_f)^c \qquad\qquad [3.2]$$

where ε_e and ε_p are the elastic and plastic deformations. The concavity of the endurance curve is supposed to be due to the plasticity of metals. According to the quasi-linear monotonic stress-strain cruve of composite materials a significant modification of the shape of the endurance curve is expected. If we assume that we can neglect ε_p, the lifetime will be given with the following equation:

$$\Delta\varepsilon_e/2 = b\,(2N_f)^b \qquad\qquad [3.3]$$

whose graph, with logarithmic coordinates, will be quite planar. This type of endurance curve is usually true for high-performance composite materials when lifetime is given as a function of amplitude of the deformation or the applied stress (see Figures 3.9 and 3.10).

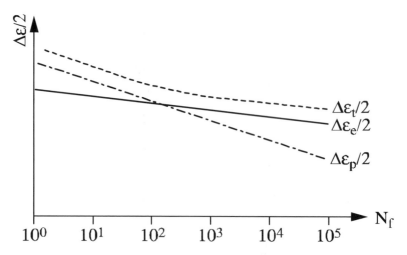

Figure 3.9. *Modeling of the endurance curve for high-performance composites from the concept of low-cycle fatigue used in metallurgy*

3.2.3. Shape of the endurance curves of composite materials

Composite materials made from unidirectional fibers (glass or carbon) totally match the hypotheses developed in section 3.2.2. The SN curve is flat and can be represented as undulating tension with the following equation:

$$\Delta\varepsilon = a - b \, (\log N_f) \qquad\qquad [3.4]$$

which was proposed in [LOR 86]. In this case, $\Delta\varepsilon$ is an almost elastic deformation (of about 1%) at the endurance limit. The experiments show that the unidirectional fiber composite materials were sensitive to an elastic low-cycle fatigue, with a sudden break of the fatigue causing damage as soon as the applied deformation becomes lower than a threshold. This gives us a better idea of the phenomenon than the asymptotic notion.

In the case of stratified composite materials with crossed layers, the fatigue damage law depends on the draping orientation. Nevertheless, the SN curve remains flat compared to metallic materials. Although it is hard to generalize, it seems that a power function allows us to represent this phenomenon, at least within the stratified materials:

$$\text{Log } S = \log F - 1/m \, \log N_f \qquad\qquad [3.5]$$

where F and m depend on the material (see Figure 3.10).

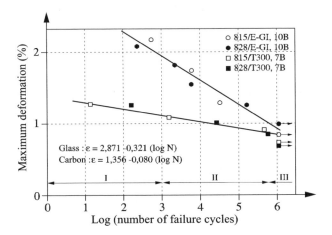

Figure 3.10. *Comparison of the endurance curves for some glass-fiber and carbon-fiber composites*

3.2.4. *Role of the fibers and the matrix*

With physical mechanisms, fatigue damage occurs as transversal cracks within the matrix and between the fibers. This is because the initiation and growth of these cracks is related to the nature of the matrix and to the orientation of the layers regarding the loading. If draping is acceptable, delamination can then occur between the layers and finally lead to fracture of the fibers. This last step depends on the nature of the fibers. This is the reason glass fibers become more damaged than carbon ones [LOR 86]. For instance, within unidirectional fiber composites, the ratio between deformation at the endurance limit and at failure under tension conditions, respectively, is:

$$\varepsilon_D/\varepsilon_F = 0.8/1.1 = 0.7, \text{ for the T 300 carbon fiber} \qquad [3.6]$$

$$\varepsilon_D/\varepsilon_F = 1.1/4 = 0.3, \text{ for the E- G1 glass fiber} \qquad [3.7]$$

These results [LOR 86] show that for the same resin, monolayer composites made of glass fiber and the ones made of carbon fiber, respectively, lead to σ_D/Rm ratios of 0.5 and 0.7. In the case of carbon fiber, it seems that increasing ε_f to 1.8% slightly increases the endurance limit of unidirectional composites with an epoxide matrix [BAR 87, CUR 87]. We can nevertheless observe some results [BAR 87] obtained with high-performance carbon fibers ST3 ($Rm = 4,300$ MPa and $\varepsilon_f = 1.8$) and a ductile epoxy resin that show an improved fatigue strength compared to regular behavior. An example is the T800 fiber. For some draping of the $[0_2/90_2]$ type, the σ_D/Rm ratio remains close to 0.8, but with an endurance limit that reaches 800 MPa for the ST3 fibers compared to 650 MPa for ordinary fibers. The endurance limit σ_D here is the maximum stress of the cycle for a loading ratio close to zero.

The fatigue strength obviously depends on the draping of the composite and the nature of the matrix. A certain number of results from the literature show that the σ_D/Rm ratio ranges from 0.4–0.8 when the arrangement of the fibers varies from orthotropic to quasi-isotropic symmetry.

Both the matrix and fabrication process have a sensitive role. We should mention that for unidirectional carbon-fiber composites, the epoxide resins are better than the thermo-plastic ones [CUR 87]. With cross-stratified materials, the ductile matrices give some higher endurance limits; e.g. resin 914 has a σ_D increase of 25% compared to resin 5208 [ROU 86]. A plausible reason is likely to be the delamination strength of the ductile resins.

3.3. Notch effect on fatigue strength

It is now well-known that metals are sensitive to notches and that their endurance limit significantly drops under fatigue with a high number of cycles. With composite materials, this phenomenon is almost unknown, which gives them a serious advantage compared to the metals as shown in Figures 3.11 and 3.12. These two figures show a key point of the behavior of composites: they are insensitive to a notch beyond 10^6 cycles and become very vulnerable to this effect below 10^5 cycles. This is the opposite effect to metals.

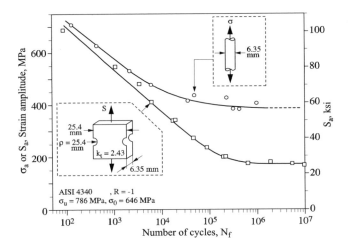

Figure 3.11. *Notch effect on the SN curve of a 4340 steel*

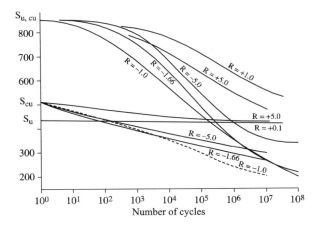

Figure 3.12. *Notch effect within a carbon fiber composite for different R ratios*

The most striking example was reported by Williams [RAM 77]. He showed that a composite plate made of carbon fiber that is quasi-isotropic and notched has an endurance limit higher than the residual strength of the unnotched material as long as we apply a loading with increasing thresholds (see Figure 3.13).

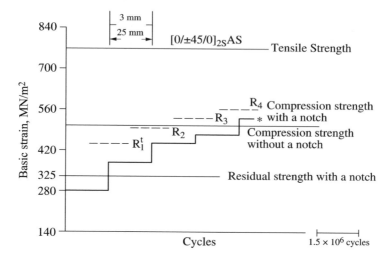

Figure 3.13. *Notch effect within an AS/3501 composite [0/+/-45/0]$_2$*

Endurance curves [ROU 86] plotted for some T 300/5208 and 914 quasi-isotropic composites show that the fatigue limit under repeated tension with 10^6 cycles only differs by 10% in the notched material compared to the initial state. As we will see much later, these results can be explained by the relaxation of the stresss due to damage.

The behavior of composite materials under fatigue that are made of fabrics is also very good when they are notched [XIA 94]. Some studies on glass/epoxy fabrics with various draping show that the ratio between the fatigue limit under tension-tension over the strength is the same with or without any notch, which comes to about 0.4 (see Figure 3.14). Therefore, the prediction of the lifetime under fatigue of a composite made of glass fiber fabrics is therefore easy to calculate from the residual strength as we have the following as a first approximation:

$$\frac{\sigma_D}{R}\text{notched} = \frac{\sigma_D}{R}\text{unnotched} \qquad [3.8]$$

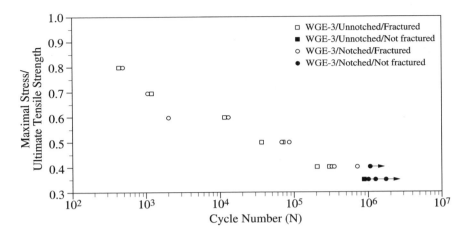

Figure 3.14. *The unique S-N curves for some unnotched composites made of glass-fiber fabrics over the residual strength*

Some other studies [BAK 86, LAG 86] have come to the same conclusion: the composite materials, although very sensitive to the notch effect under a monotonic stress, are not at all sensitive under a cyclic stress. We can compare the notch effect within some specimens made from a 7075 T6 aluminum alloy and quasi-isotropic carbon-fiber composite presenting a notch with $K_t = 3.1$. Here, we find that the stress concentration, K_f, under fatigue is <1 for the composite and equal to 2.5 for the alloy, for a lifetime, N_f, of 10^7 cycles [SCH 87]. We still have to determine whether we can generalize this result for any fatigue loading, in particular in the case where part of the cycle involves compression.

3.4. Effect of a stress on composite fatigue

3.4.1. *Fatigue under compression*

The influence of ratio, R, between the minimum stress of the cycle and the maximum stress does not have the same effect within the metals and composites.

3.4.1.1. *Fatigue of unnotched sheets*

Under monotonic stress, all composite materials present a compression strength lower than the traction strength. This is given as a function of the reinforcement: boron, carbon, glass, Kevlar; the strength decreasing according to the order of the list given.

The same applies to unnotched composites under cyclic stress. At the limit, when the fatigue cycle is entirely under compression, fracture can occur. Different studies [RAM 77, SCH 87] have led to the conclusion that the fatigue limit of composites is lower under cyclic stress than when the stress is entirely compressive ($R = 10$) or the material is under tension-tension conditions ($R = 0.1$), see Figure 3.15.

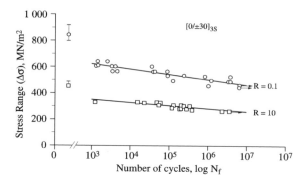

Figure 3.15. *SN curve for a carbon-fiber composite under compression and tension conditions*

This last type of damage is unknown in the case of metals and alloys that do not break under compressive fatigue.

Knowing that composite materials are not sensitive to the notch effect under undulating tension but that they can be more rapidly damaged under alternate tension, it is interesting to consider the compressive fatigue of notched composite sheets.

3.4.1.2. *Fatigue of notched sheets under compression*

To study the behavior of composite materials under compression in greater detail, we propose to use a modified compact specimen. This will allow us to avoid generalized buckling of the stratified sheet when the compressive force is significant. The compressive fatigue test with modified compact specimens is also described [LAI 87, SCH 87].

For instance, the case of the notched T300/5208 composite is presented. We should first mention that fracture loading of the same quasi-isotropic T300/5208 composite specimen is more important under compression than under tension. This means that, on a first approximation, the residual tensile strength is lower [SCH 87]. Results show that for a cyclic stress under compression-compression conditions, $R = 10$, the σ_D/R_m ratio under compression tending towards 0.4, whereas it is >0.5 under tension-tension conditions. In the intermediate case, where the stress is a symmetrical and alternate tensile one, the σ_D/R_m ratio is close to 0.3, compared to the monotonic tensile strength, but close to 0.5 compared to the compressive monotonous fracture.

From these observations, we can find very different behaviors within composite materials made from a carbon fiber and epoxide matrix, whether there is a notch or not and whether the stress is tensile or compressive. We can sum up the situation as follows:

– for the unnotched composite material, the endurance limit under undulating traction ranges from 0.5 UTS and UTS. It falls below 0.5 UTS under symmetrical tension compression;

– for the notched material, the endurance limit is roughly equal to the statistical residual strength under undulating tension. However, if the stress is compressive, the endurance limit is clearly lower than the compressive residual strength. Moreover, as soon as there is a cyclic compression, the endurance limit depends slightly on the presence of a notch;

– finally, whether the material is notched or not, only a compressive stress can lead to failure due to fatigue. The mechanism of the damaging process due to fatigue under compression conditions is due to the formation of delaminations within the external layers of the composite. These can then spread within the material until failure occurs. The draping of the composite, nature of the resin and fibers then have a major role and can modify the conclusions that we have just drawn.

3.4.2. *Fatigue under bending conditions*

In practice a certain number of mechanical pieces, such as springboards, are subjected to cyclic bending. When carrying out experiments it is always cheaper to carry out a fatigue test under bending than under traction conditions. These reasons lead us to focus on fatigue under bending conditions.

The analysis of the phenomenon and results is complex, as several types of damage can simultaneously occur under bending conditions: tension, shearing and compression. The lifetime of a sample subjected to bending therefore depends on its section and especially its thickness [ROO 87]. For instance, the endurance limit of a unidirectional glass-epoxy composite goes from 250 to <200 MPa when the thickness of the composite increases from 4 to 8 mm (see Figure 3.16).

In addition to this, the type of stress can influence the variation in stiffness as a function of the number of cycles. In other words, as soon as the first damage occurs, the composite material presents a different behavior.

We also have to consider the significant influence of the distance between the supporting points, which modifies the distribution of the tensile and shearing stresss within the material and leads to different delaminations. The fatigue test under bending conditions is therefore worth studying [TRO 88].

Modeling of the phenomenon is difficult as there are many parameters. Currently, the most interesting models are those obtained by predicting the drop in stiffness using Katchanov's damage theory [SID 87, TRO 88]. The distribution of the results and the number of coefficients to be determined make the method quite complex.

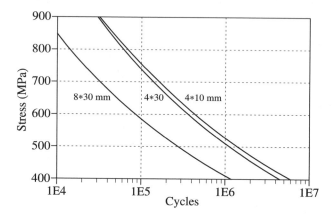

Figure 3.16. *Prediction of the lifetime of a unidirectional glass-fiber composite under three-point bending, when R = 0.1, in the case of rods of different thicknesses*

Finally, we have to consider that the distance between the supporting points influences the distribution of the tensile and shearing stresss within the material and leads to different delaminations. The fatigue test under bending conditions is therefore worth studying regarding normalization [TRO 88].

Under bending conditions, the matrix has a significant role, especially in the initiation of the delamination. Within a composite reinforced by AS4 carbon fibers, the epoxide resin is stronger than the thermoplastic one, in stroke control [CRO 87]. This result can be explained by the fact that the thermoplastic resin withstands delamination but not compression.

Overall, it is quite clear that it is not easy to compare fatigue results under fatigue conditions with results under tension or compression conditions. The behavior of composite materials under bending is therefore slightly more difficult to determine than the behavior of metals and is, in any case, very different.

3.4.3. *Effect of tensile over-loading*

A small number of speculations deal with spectrum loading and damage accumulation regarding the fatigue of composite materials. The absence of plasticity and resulting residual stresses should explain the fact that there is no retardation effect once tensile over-loading has occurred, where long crack propagation is concerned.

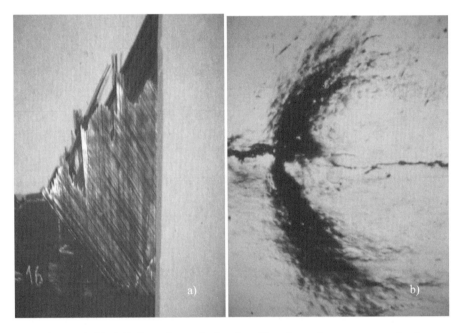

Figure 3.17. *a) Failure of a notched carbon-fiber sheet after a severe overload; and b) tacking of a crack by the same over-load in a 2024 aluminum sheet*

Figures 3.17 and 3.18 give examples of the over-loading effects of delamination of a glass-fiber composite material and the cracking of an aluminum alloy. The results in these figures imply that turbulence is beneficial for an aircraft component made in aluminum alloy but is detrimental for a composite material part. In the worst case, bending or compressive stresss associated with tensile over-loadings in a component made of composites demand much greater attention from the engineer than the metallic solution. Once more, the cumulative fatigue damage within composites is very different from that observed within metals.

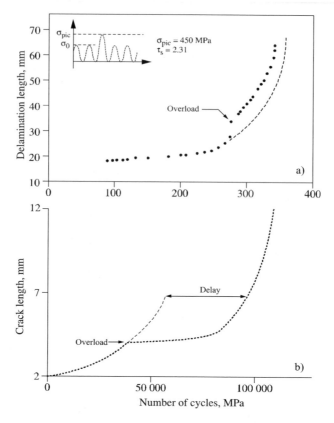

Figure 3.18. *Effect of an overload on fatigue crack: a) acceleration in composite; and b) retardation in metal*

3.5. Fatigue after impact

We cannot talk about the fatigue of composite materials if we do not consider the damaging effect of low energy impacts on the initiation of fatigue failure. Overall, the role of impacts on the fatigue of composites can be compared to the role of machining scratches on the surface of metallic materials. In particular, low energy shocks lead to delaminations within the composites, which can propagate under cyclic loading. This phenomenon is of concern as no surface damage occurs on the impacted composite. We should mention that a low-energy impact on a quasi-isotropic carbon-fiber T300/5208 composite plate leads to about a 30% decrease in UTS and a 60% decrease in compressive strength [GUN 85, WAL 77]. In other words, the effect of an impact in this case is equivalent to a hole 3 mm in diameter under traction, and 30 mm under compression. This again points out the relative weakness of these materials under compressive conditions. The fatigue limit under undulated tension of the impacted material is lower than its tensile strength [LAI 87], but under compression it is mainly the fatigue limit that drops to 20% of the tensile nominal strength of the undamaged materials (see Figure 3.19).

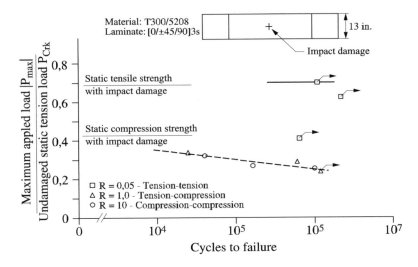

Figure 3.19. *The effect of impact on a carbon fiber composite under cyclic tension or compression conditions on the SN curve*

The damage that is due to a light impact should then be considered in the prediction of fatigue failure, as a correction to its tensile stress and as a main parameter of a compressive stress ($R = -1$ or $R = 10$). We should also mention that no precise correlation could be observed between the dimension of the impacted zone on the residual strength [GUN 85].

3.6. Fatigue damage criteria

3.6.1. *Variation of rigidity*

The growth of fatigue damage within the composite material can be studied by the evolution of its rigidity. When the stress is tensile, the rigidity varies from 10–30% before failure occurs [DAN 87, SCE 87]. We usually distinguish three phases for fatigue damaging.

A first short phase, which does not exceed 20% of the number of failure cycles, N_f, during which the first transversal cracks will be formed within the matrix.

A second phase, the longest one, which corresponds to multiplication of the transversal cracks to reach, according to Reifsnider [REI 87], a critical density called CDS.

Finally, the last phase, of about 10% of N_f. This corresponds to the final damage due to longitudinal cracks, delamination and the failure of some fibers. During the intermediate phase, the rigidity slightly varies according to a plateau [DAN 87, SCE 87], see Figure 3.20. With a notch the phenomena remain the same, but they are obviously localized around the notch, which leads to a reduced variation in rigidity. The rigidity then becomes more difficult to observe within notched plates subjected to cyclic compression.

A was study carried out on T 300/5208 composites. This showed that with compliance variation under compression conditions, despite delaminations close to the surface, it is necessary to measure the rigidity in transversal deformation (ε_{33}) in order to study the damaging process [SCH 87].

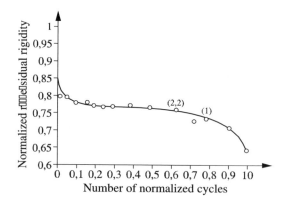

Figure 3.20. *Evolution of the rigidity of a composite subjected a tensile cyclic stress*

Figure 3.21 illustrates the three-stage phenomenon of fatigue damage:

– the first stage related to the occurrence of the first delamination close to the surface;

– the second stage depending on growth of the delamination under the compression effect; and

– the third stage, related to the multiplication of in-depth delaminations [LAI 87].

It is always better to consider the stiffness loss as a good damage criterion. However, this is not the case with residual strength after fatigue, which follows laws that are more complex.

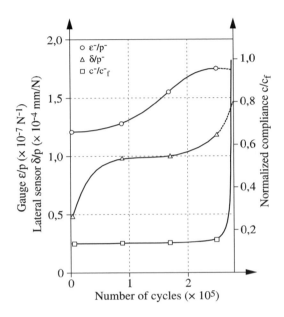

Figure 3.21. *Evolution of compliance and local deformation of a notched carbon-fiber composite under compression-compression loading*

3.6.2. *Variation of residual strength after fatigue*

The variation of residual strength after fatigue is not a reliable indicator of the damage process. Thus, it is not that rare to measure an increase in residual strength from 10–40% in carbon-fiber composites after a fraction of the number of cycles below 50% of the lifetime of the sample [BAK 86, RAM 77] for any cyclic stress (see Figure 3.22).

This improvement in composite residual strength can frequently be observed when there is a notch, as fatigue creates microcracks or delaminations, which relax the stress concentration around the notch. Nevertheless, despite the improvement in residual strength, rigidity drops.

We should also note that the residual compression strength is affected to a small extent by the preliminary compressive cycle [BAK 86].

All this shows that this parameter is not really adapted to measure fatigue, even if the variation in residual strength is very interesting from a practical perspective.

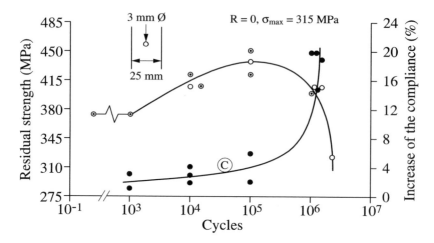

Figure 3.22. *Evolution of the residual strength and compliance of T300/934 composites subjected to undulated tensile fatigue*

3.7. Conclusion

The fatigue of composite materials made from polymers reinforced by long fibers is different in several ways from the fatigue of metallic materials. We can sum up these differences as follows:

– the specific endurance strength of composite materials subjected to cyclic tensile loading is generally greater than the strength of metals;

– composite materials are less sensitive under fatigue to the notch effect compared to metals;

– cyclic compressive loads lead to the serious damage of composite materials;

– the damage mechanisms of composites do not only evolve at the surface, as is the case with metals, but they also do so within the material;

– over-loading tensile effects are positive to the propagation of damage within the metals but do not affect composites.

In summary, the fatigue of composite materials depends on the nature of the fibers and resins, on the draping and on the quality of the interfaces. There are many parameters that have to be optimized to improve fatigue strength.

3.8. Bibliography

[BAK 86] C.E. BAKIS, W.W. STINCHCOMB, "Response of thick, notched laminates subjected to tension-compression cyclic loads", *STP 906*, pp. 314-334, 1986.

[BAR 87] C. BARON, K. SCHULTE, "Fatigue damage response of CFRP with toughened matrices and improved fibres", *ICCM VI*, pp. 265-275, 1987.

[BAT 04] C. BATHIAS, P. PARIS, *Gigacycle Fatigue in Mechanical Practice*, Dekker, New York, 2005.

[BAT 05] C. BATHIAS, *Matériaux Composites*, Dunod, Paris, 2003.

[CAR 86] L. CARLSSON, C. EIDEFELDT, T. MOHLIN, "Influence of sublaminate cracks on the tension fatigue behaviour of a graphite-epoxy laminate", *STP 906*, pp. 361-384, 1986.

[CHA 86] A. CHAREWICZ, I,M. DANIEL, "Damage mechanisms and accumulation in graphite/epoxy laminates", *STP 907*, pp. 274-297, 1986.

[CRO 87] R.B. CROMAN, "Flex fatigue of AS4 graphite reinforced thermoplastics", *ICCM VI*, pp. 476-488, 1987.

[CUR 87] P.T. CURTIS, "An investigation of the tensile fatigue behaviour of improved carbon fibre composite materials", *ICCM VI*, pp. 454-455, 1987.

[DAN 87] I.M. DANIEL, J.W. LEE, G. YANIV, "Damage mechanisms and stiffness degradation in graphite-epoxy composites", *ICCM VI*, pp. 4129-4138, 1987.

[GUN 85] E.G. GUYN, T.K. O'BRIEN, "The influence of lay-up and thickness on composite impact damage and compression strength", *AIAA, 85-0646*, pp. 187-196, 1985.

[HAR 03] B. HARRIS, "Fatigue in Composites", *CRC/WP 2003/3*, Cambridge, 2003.

[KEL 87] S. KELLAS, J. MORTON, S.M. BISHOP, "The effect of environment on the fatigue damage development in notched carbon fibre composites", *ICCM VI*, 4139-4149, 1987.

[LAD 03] P. LADEVEZE, G. LUBINEAU, "A computational meso-damage model for life prediction for laminates", *CRC/MP/432-441*, 2003.

[LAG 86] P.A. LAGACE, S.C. NOLET, "Effect of ply thickness on longitudinal splitting and delamination in graphite epoxy under compressive cyclic loading', *STP 906*, pp. 335-360, 1986.

[LAI 87] D. LAI, C. BATHIAS, "The compression fatigue resistance of a carbon-epoxy composite plate containing a hole", *ICM 5*, pp. 1231-1238, 1987.

[LOR 86] L. LORENZO, H.T. HAHN, "Fatigue failure mechanisms in unidirectional composites", *STP 907*, pp. 210-232, 1986.

[OGI 02] S. OGIHARA, N, TAKEDA, S, KOBAYASHI, "Damage mechanics characterization of transverse cracking behaviour in quasi isotropic CFRP laminates with interlaminar toughened layers", *Int. Jour. of Fatigue 24*, pp. 98-107, 2002.

[RAM 77] S.V. RAMANI, J.G. WILLIAMS, "Notched and unnotched fatigue behaviour of angle ply graphite/epoxy composites", *STP 638*, pp. 27-46, 1977.

[REI 87] K.L. REIFSNIDER, "Life prediction analysis", *ICCM VI*, pp. 41-431, 1987.

[ROO 87] P. DE ROO, B. PALUCH, "Etude expérimentale et modélisation de l'endommagement en fatigue d'un composite unidirectionnel", *ICCM VI*, pp. 4431-4442, 1987.

[ROU 86] M. ROUCHON, M. LIBERGE, Endommagement des Matériaux Composites Carbone-epoxy, CEAT Report, 5213, CEAT, 1986.

[SCH 87a] K. SCHULTE, "Fatigue behavior and damage development in woven fabric and hybrid fabric composites", *ICCM VI*, pp. 489-499, 1987.

[SCH 87b] D. SCHUTZ, J. GERHARZ, "Fatigue strength of a fibre reinforced materials, Composites", *ICCM VI*, pp. 245-250, 1987.

[SID 87] F. SIDOROFF, B. SUBAGIO, "Fatigue damage modeling of composite materials from bending tests", *ICCM VI*, pp. 432-439, 1987.

[TRO 88] J.P. TROTIGNON, L. VINCENT, *Fatigue of Polymer Composites*, Vamas report, Vamas, 1988.

[WAL 77] R.W. WALTER, "Designing for integrity in long-life composite aircraft structures", *STP 636*, pp. 228-247, 1977.

[WHI 74] J.M. WHITNEY, R.J. NUISMER, "Stress fractrure criteria for laminated composite containing stress concentration", *Jour. Comp. Mat.*, pp. 253-256, 1974.

[XIA 94] J. XIAO, C. BATHIAS, "Fatigue damage and fracture mechanism of notched woven laminates", *Composite Materials*, vol. 28, no. 12, pp. 1127-1139, 1994.

Chapter 4

Fatigue of Polymers and Elastomers

4.1. Introduction

As Hertzberg said in his book [HER 80], the fatigue of polymers is a relatively recent issue that goes back to the second half of the 20th century, with the notable introduction of plastics into mechanical pieces. It obviously does not matter whether a plastic toy loses its resistance and becomes cracked due to fatigue. However, it is entirely different when we are dealing with an automotive component: engine, timing belt or tires.

The fatigue of polymers is presented in the same way as the fatigue of metals. The initiation of cracks is different from their propagation. We must still bear in mind that this very Cartesian approach hides some very different damaging mechanisms. Polymers are not subjected to plastic deformation as metals are. In addition, some polymers are amorphous, whereas others are partly crystallized.

In this chapter, research in this domain will be presented by developing one of the most typical aspects of the fatigue of polymers at large deformation in elastomers. Elastomers are interesting from both technological and fundamental points of view. Their deformation behavior coupled with the elastic state perfectly illustrates this topic, whereas polymers said to be "fragile" present a mechanical behavior closer to the behavior of metals.

Chapter written by Claude BATHIAS.

4.2. Life of polymers

Since the first studies by Dillon [DIL 50] and Rosen [ROS 64], the lifetime of polymers under fatigue is estimated using a version of the Wöhler curve. As polymers are always affected by viscoelasticity, the lifetime under fatigue does not only depend on the mean stress of the cycle and the alternate stress but also on:

– the frequency;

– shape of the cycle;

– temperature.

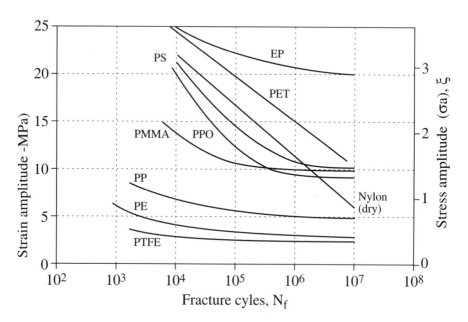

Figure 4.1. *S-N curves for different polymers at 30 Hz*

The study of the endurance of polymers leads to the selection of materials to be used for various purposes, as indicated in Figure 4.1. Nevertheless, a frequency change in the fatigue test can lead to a significant change in lifetime. An increase in frequency leads to a drastic reduction in the number of cycles due to the rise in temperature.

We should mention that the heating dissipation of the polymer is a consequence of the combined effect of the frequency and amplitude of the stress, which can explain that at a low frequency, of about 1 Hz, two mechanisms of damage are involved: mechanical damage and thermal damage as a function of the strain level. The Wöhler curve then becomes sigmoid (see Figure 4.2).

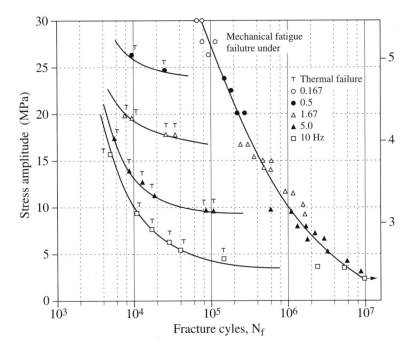

Figure 4.2. *Thermal fatigue and traction-compression mechanical fatigue*

We can also observe that polymer fatigue tests are carried out under tension, compression, torsion conditions and, unfortunately (most of the time), under bending conditions for economic reasons. One of the fatigue test rules under bending is not scientifically satisfying as it leads to several types of damage that cannot easily be distinguished. For composite materials, axial tests are favored, even if they are quite costly.

4.3. Crack propagation within polymers

Failure mechanics is applied to the propagation of cracks in polymers by adopting Paris' law. To consider viscoelasticity, we can write:

$$\frac{da}{dN} = CK_{average}^{m'} \rightleftharpoons [\Delta K]^m \rightleftharpoons f^n \qquad\qquad [4.1]$$

where C, m', m and n are some constants of the material and $K_{average}$ the average stress intensity factor. Or we can have [HER 80]:

$$\frac{da}{dN} = CK_{average}[\Delta K] \qquad\qquad [4.2]$$

Finally, Moet proposed calculating the crack growth from the mechanical hysteresis energy, W_i:

$$\frac{da}{dN} = \frac{\beta W_i}{\gamma R - J} \qquad\qquad [4.3]$$

where γR is the energy required for the damaged zone at the crack tip and J is the dissipated energy rate.

In reality, the application of the failure mechanics to the cracking of polymers is much less easy than it is for metals. Figure 4.3 shows the results of crack propagation given as a function of ΔK for various polymers. This leads to a selection, or possibly the determination of a propagation threshold, but certainly not to a unique equation between log da/dN and log (ΔK).

This approach, which is based on the fracture mechanics and configured with K, is limited to "fragile" polymers. In the case of very ductile polymers and a fortiori to elastomers, the notion of stress intensity factor is no longer suitable. The Griffith energetic criterion, G, is thus used. This is considered by polymerists to be equal to the tearing energy, T, whose definition is the same as the Griffith criterion.

4.4. Damaging mechanisms of polymers

Depending on the alternate stress and on the frequency, two main mechanisms can explain the cyclic damaging of polymers: the first is mechanical and the second is of thermal origin.

Due to the viscoelastic state, stress and deformation are not in phase during cycling. This phase difference, as well as the heat dissipation related to mechanical hysteresis, both depend on the frequency. When the increase in temperature becomes significant, premature failure occurs due to the thermal instability of the polymer (see Figure 4.2).

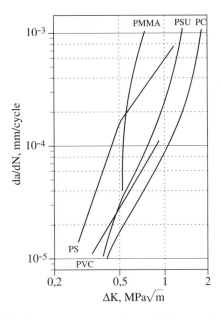

Figure 4.3. *Propagation curves of cracks for several polymers*

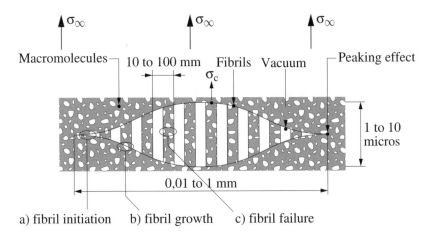

Figure 4.4. *Microstructure of a craze without any crack*

The fundamental damaging mechanism of polymers comes from the formation of crazes (see Figure 4.4). The notion of crazes deals with the formation of microcracks, but also with the orientation of the macromolecules, depending on the stress field. When crack propagation occurs, crazes are formed before the crack tip (see Figure 4.5). The formation of crazes depends on the stress, frequency and transition temperature, T_c, of polymers. Crazes are typically due to low stresses and to the lowest temperatures [ROS 64].

However, for higher stresses, damage at the crack tip moves towards both maximum shearing bands (see Figure 4.5), oriented at 45° to axial loading. What happens here is similar to the plan strain states within metals.

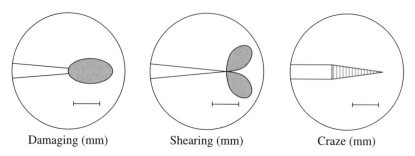

Damaging (mm)   Shearing (mm)   Craze (mm)

Figure 4.5. *Different types of damage at the crack tip*

In addition, fatigue cracking of polymers is strongly influenced by the chemical damage due to the environment. Oxygen, and sometime ozone, can lead to the failure of molecular chains due to oxidation. The presence of liquid, mainly water, contributes to the transformation or dissolution of the additives of the polymers and thus to the modification of the properties. Needless to say, the chemical composition of the polymers – the molecular weight – influences the fatigue strength of polymers. Finally, the propagation mechanism of fatigue cracks is sometimes observed due to cleavages that are formed on the fracture surface under high loading.

4.5. Specific case of the fatigue of elastomers

As soon as rubber is subjected to a cyclic stress, its temperature increases, especially within the material. This temperature increase results in a modification of the properties and possibly damage. Within natural rubber and some synthetic rubbers, the amorphous microstructure tends to partly crystallize when the deformation is significant. In addition to this, when under very low contact pressure it seems that rubber can be subject to chemical damage along with a solid-liquid phase change. This may also be due to the influence of the ambient air. In other words, under natural stress conditions rubber fatigue is a function of the combination of three types of damage: mechanical, thermal and chemical damage.

Experiment shows that under cyclic stress, rubbers become damaged until the formation of one or several main cracks that propagate. As in metals, we usually study crack initiation and crack propagation separately. As failures of service pieces usually occur due to manufacturing or accidental defects, understanding the growth of cracks is really important. This way we can select rubbers with the best fatigue characteristics and correctly design rubber pieces.

Rivlin and Thomas [RIV 53] in the 1950s, then Lake [LAK 83] and Lindley [LIN 73] in the 1960s, and finally Gent and Stevenson in the 1980s [GEN 92, STE 83] applied the Griffith criterion (G) to the tearing of elastomers. In this case, these authors replaced the level of energy dissipation (G) with the notion of tearing energy, T, whose physical definition is similar.

Some issues can occur with the application of fracture mechanics to rubbers due to the deformable property of elastomers. In particular, the elasticity of elastomers is nonlinear and cracks subjected to high deformations do not remain sharp, as the Griffith model suggests. Nevertheless, energy does not dissipate due to plasticity and the failure mechanisms of rubbers are of the "fragile" type. When failure mechanics are applied to fatigue cracking, we usually have to keep in mind that cyclic damage of rubbers depends on mechanical stress (frequency, ratio R), temperature, thermal dissipation and finally on the environment and the combination of all these parameters. Atmospheric oxygen is known to significantly influence rubber fatigue. In what follows, a comprehensive approach to the fatigue cracking of natural rubbers is proposed by defining the main mechanical parameters and influence of the mechanisms of damage related to the microstructural aspect of rubbers, in order to understand crack propagation.

4.6. The life of natural rubbers

Natural rubbers reinforced with carbon black are amorphous without stretching at ambient temperature. Stretched by a traction loading, these rubbers partly crystallize. As the mechanical properties are clearly different before and after crystallization, it is expected that the mean stress will have a strong influence on the fatigue life.

In Figure 4.7 the effect of the mean stress on lifespan is illustrated by some results obtained on a natural rubber subjected to an alternate stress of 1 MPa and then of 1.5 MPa. Specimens are axi-symmetrical, with an hourglass shape 26 mm in diameter (see Figure 4.6).

For any alternate amplitude level, the curves have the same shape. They all present a knee corresponding to an alternate stress value that is almost equal to the alternate stress applied. For mean stresses that are lower than the alternate stress value, corresponding to a traction-compression mode, fatigue life decreases when the stress increases. This is usually the case of a Wöhler curve.

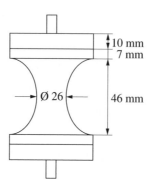

Figure 4.6. *Endurance specimen*

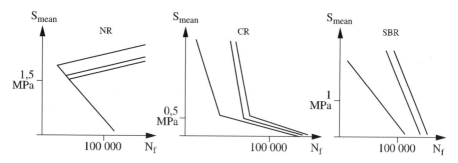

Figure 4.7. *Effect of mean stress on the fatigue life of rubbers*
(NR – natural rubber; CF – polychloroprene; SBR – polybutadiene)

Under a traction-traction mode (mean stress higher than the applied alternate strain), the mechanism is suddenly reversed: the rubber becomes reinforced and has better behavior under increasing mean stress. This reinforcement is due to the formation of crystallite, which prevents any crack initiation and progression from occurring.

4.6.1. *Wöhler law with crystallizable rubbers*

To characterize the behavior of materials under fatigue conditions, Wöhler's representation that defines the stress-number of cycle equation is generally used:

$$\text{Log } N_f = a - b\,\sigma \qquad\qquad [4.4]$$

In this representation, a and b are two constants and σ is usually the amplitude of the applied stress. This equation can be represented by a curve with $\sigma - \text{Log } N_f$ coordinates.

As the compression stress does not mechanically damage rubber, its role is to destroy the crystallites that were formed under traction. For this reason, it would be more suitable to give Wöhler's equation with the maximum stress instead of the stress amplitude. The equation can then be written as:

$$\text{Log } N_f = a - b\,\sigma_{max} \qquad\qquad [4.5]$$

In order to apply the Wöhler equation, a stress term σ_{crys} is introduced that comes from the crystallization effect. The equation then becomes:

$$\text{Log } N_f = a - b\,(\sigma_{max} - \sigma_{crys}) \qquad\qquad [4.6]$$

$$\sigma_{crys} = 0 \text{ when } \sigma_{min} < 0 \text{ and } \sigma_{crys} > 0 \text{ when } \sigma_{min} > 0 \qquad [4.7]$$

4.7. Crack propagation in natural rubber

In order to apply a traction or compression loading, a thick specimen bearing a lateral notch should be used. In order to avoid the geometry effect, large specimens (see Figure 4.8) with a length of 150 mm are used. The fatigue crack is initiated from a mechanical notch cut with a razor blade with a length of 20 mm. These specimens can be tested in tension or in compression. Crack propagation is studied as a function of the dissipated energy rate, given by the following equation:

$$T = \frac{1}{b}\frac{dU}{dA} \qquad\qquad [4.8]$$

where U is the elastic energy for a crack with a surface of A. Thickness b is corrected by the lateral deformation at the crack tip when T is determined thanks to the compliance method. Crack length a is measured with a video camera recorder. According to Griffith's criterion, the dissipated energy level is calculated in the case of crack extension when failure occurs. The growth of the crack then comes to:

$$\Delta T = T_{max} - T_{min}$$

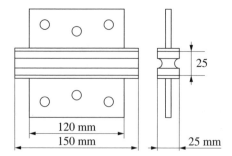

Figure 4.8. *Crack growth specimen to be used for rubbers*

4.7.1. *Crack growth curve – influence of ratio R*

Cracking experiments confirm the existence of a sigmoid relationship between da/dN and ΔT, which can be written as:

$$\frac{da}{dN} = C.\Delta T^n \qquad [4.9]$$

For low values of ΔT, a non-cracking threshold (ΔTs) can be defined (see Figure 4.9). The threshold significantly increases when ratio R increases, which is different to what is observed in metals. We should point out that when ratio R reaches 0.5, the crack no longer propagates under fatigue conditions, whatever the maximum force applied.

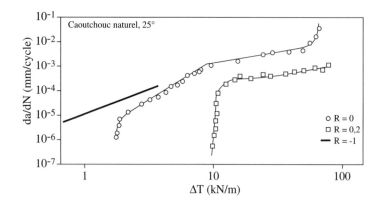

Figure 4.9. *Crack growth curve as a function of ratio R*

When ratio R reaches -1, under symmetrical traction-compression conditions, the ΔTs threshold vanishes. As a consequence, the fatigue crack does not stop growing, even for the lowest stress when $R = -1$.

When we have a traction-compression strain, a chemical decomposition of the rubber can be observed at the crack tip where black drops appear. The cyclic damage under compression is likely to lead to oxidation, a chemical reaction, which occurs after mechanical damage.

In the reverse case, where ratio R is equal to 0.5, elongation at the crack tip is of several hundred percent, which leads to local crystallization of the rubber, which goes against crack propagation. This crystallization effect stops occurring when $R = -1$, as the reversibility of crystallization occurs during the compression phase.

We conclude that the propagation strength of fatigue cracks in natural rubber is much higher when the static stress of the cycle is high. Also, fatigue propagation is faster at the point of minimum stress in the cycle under compression.

4.7.2. Influence of the test temperature

As we might expect, the crack growth rate increases when the temperature increases (see Figure 4.10). For a ratio $R = 0.2$, the cracking threshold drops from 10 to 3 kN/m between 25°C and 80°C, as the tests were carried out in air. This weakening of strength with temperature is related to the increase in mechanical and chemical damage as well as the decrease in crystallization.

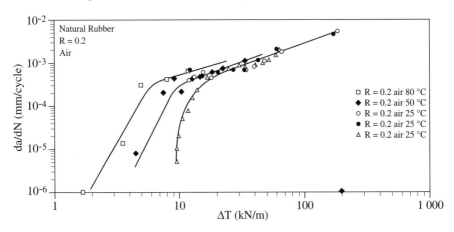

Figure 4.10. *Influence of temperature on natural rubber fatigue crack growth*

4.7.3. *Influence of the environment and combined environment-temperature effects*

The environment obviously influences chemical damage. Some tests carried out in air, water and under dry nitrogen at 25°C clearly show the effect of gaseous oxygen (see Figures 4.11 and 4.12). In water, the fatigue crack threshold is higher than in air. The dry nitrogen environment gives the highest threshold. The three thresholds are 10, 20 and 50 kN/m, respectively, when $R = 0.2$.

To make sure that the effect of water is not related to thermal dissipation, some tests run in water and in air at 50°C have been compared. The test results show that the cracking threshold at 50°C is still higher in water than in air. Thus, the environment effect occurs in the same way at 25°C and 50°C. This study shows that gaseous oxygen accelerates fatigue cracking at room temperature and also at higher temperatures when ratio R is positive. This means that crystallization occurs as soon as the crack is opened. When crystallization does not occur at $R = -1$, it seems that fatigue cracking is the same in air and in water. To get a fatigue crack threshold when $R = -1$, all traces of oxygen by cracking have been removed under pure nitrogen. The threshold then reaches 10 kN/m (see Figure 4.11).

It is concluded from this experiment that the influence of the environment is more significant when the amorphous transformation is inhibited by compression at the crack tip. Under traction-compression conditions, oxygen traces can keep the chemical damage going at the crack tip. This only disappears under a nitrogen atmosphere.

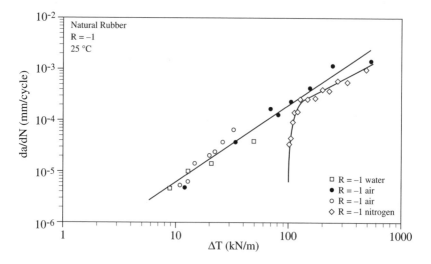

Figure 4.11. *Influence of the environment on natural rubber cracking*

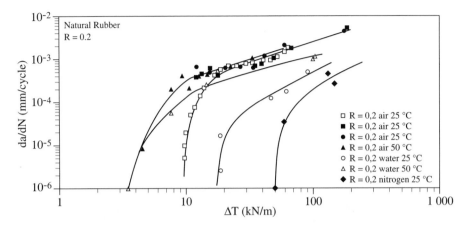

Figure 4.12. *Combined influence of the temperature and environment*

4.8. Propagation mechanisms of cracks in natural rubber

Different cracking mechanisms due to fatigue were observed by fractography. These mechanisms are specific and distinct from those observed in other polymers.

When the crack growth rates are low (at the cracking threshold), the fracture surface becomes fragile. Secondary microcracks are multiplied. For faster rates, we can observe more plane fractures and a rougher relief. The existence of fatigue striations can be observed for high crack growth rates (10^{-3} mm/cycle); see Figure 4.13.

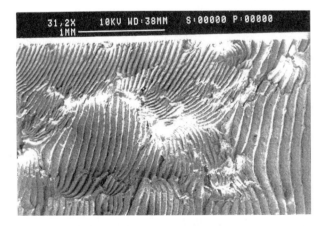

Figure 4.13. *Striations on the fracture surface of a natural rubber*

Fracture facies obtained when $R = -1$ present small transformed zones due to the high compression levels obtained, leading to a significant chemical deterioration. The higher ratio R is, the more wrenching we can observe. Nevertheless, the distribution of striations is more homogeneous. We can also see the presence of many small tongues of about 1 µm.

Whatever the loading ratio, at high magnification we can find a typical structure showing the failure of rubber. This structure involves some very elongated strips folded in on themselves, which have undergone an irreversible deformation. The size of these strips or tongues is of a few microns and they can be found on the surface of the entire specimen (see Figure 4.14).

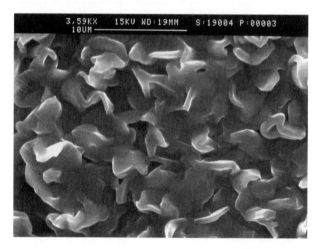

Figure 4.14. *Strips or tongues at high magnification*

Finally, whatever ratio R is studied, the reinforcement phenomenon due to crystallization of stretched chains does not directly influence the microstructure of the fracture facies. We should note that when the temperature increases, more strips are formed and their shape gets better. We do not find any strips in water or under nitrogen; mediums where there is no gaseous oxygen. As a consequence, these strips are typical of chemical damage due to gaseous oxygen of the amorphous rubber, with a possiblly more pronounced deterioration due to the increase in temperature.

The study of the damage mechanisms of natural rubber cracking leads to the following conclusions:

– pre-strain under traction conditions improves the behavior under fatigue thanks to the crystallization of the stretched chains. However, switching to compression causes much more damage to the material;

– in addition to mechanical damage, significant chemical damage has to be considered due to the oxygen in air. Moreover, when the temperature increases, the oxidation reaction is accelerated. It is therefore worth working under an inert atmosphere, such as nitrogen, for any temperature. A non-cracking threshold can be obtained when $R = -1$, whereas it did not exist in water or air;

– finally, if the fractographic study does not allow us to totally link the microstructure to the cracking mechanisms of this rubber, it still allows us to qualitatively observe the influence of the various factors related to damage. The microscopic formation of the strips strongly depends on the gaseous oxygen and the amorphous property of natural rubber.

The example of natural rubber clearly shows the interdependence of the mechanisms of fatigue damage of polymers: mechanical stress, chemical activity, the effect of temperature and distribution of the macromolecules.

4.9. Multiaxial fatigue of rubbers

Loading, usually occurring within parts and structures made from elastomer, is multidirectional. In the case of metals subjected to multiaxial fatigue, the stress tensor deviator is the principal parameter related to the fatigue life, as the damaging of metals is mainly due to plastic deformation. The plasticity of metals depends on the sliding of the dislocations and this depends on the shear due to the stress tensor deviator. This is why the equivalent von Mises stress is successfully used to define fatigue life under multiaxial fatigue. Can we apply this approach to elastomers? The answer to this question is no, as there is no plasticity in elastomers according to Friedel's dislocations theory.

Determining and testing the fatigue damage of rubbers is different from that with metals. Bathias and Le Gorju [BAT 02] previously showed that, within rubber, shear leads to quasi-cleavages; this not the case in metals. They also showed that the stress becomes more biaxial and thus makes ductility stronger, leading to some mechanisms of pseudo-dimple formation. Finally, they showed that due to a given hydrostatic tension, cavitations are formed on rubber. All these mechanisms operate under biaxial fatigue. They are obviously different from the mechanisms observed within metals. We do not see how, in these conditions, von Mises stress can correctly define the fatigue of rubbers. For the same reasons, the prediction of the multiaxial fatigue of rubbers cannot be established based on a single parameter.

Nevertheless, it seems that the maximum principal stress usually controls damaging under biaxial fatigue conditions. This is why some recent publications do not focus on the fundamental data on the fatigue of elastomers that cannot be presented by a single parameter approach.

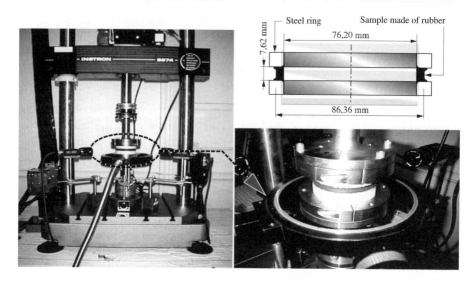

Figure 4.15. *Equipment for determining the biaxial fatigue of elastomers*

Figure 4.16 gives an example of a prediction of the fatigue life of a natural rubber as a function of the main strain for some traction-compression-torsion tests.

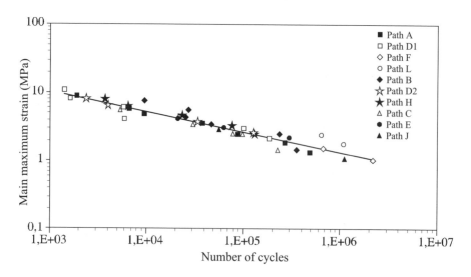

Figure 4.16. *SN curve of a natural rubber under traction-torsion conditions*

4.10. Cavitation of rubbers

Cavitation is a peculiar damage phenomenon that is specific to rubbers. When a cylindrical and axisymmetrical specimen (planar deformation) is strained, some cavitations are formed in its core. These grow and multiply with the elongation or number of cycles and progressively lead to failure of the material.

These cavities are not holes or bubbles. They are due to the formation of a strong triaxial stress field within the material that leads to spherical conformations of the macromolecules when the hydrostatic stress (average of the main stress) reaches a certain critical value.

The cavitation property cannot be given according to a critical elongation. However, the numerical calculation linked to non-destructive control by tomography allowed us to show that the cavitation of natural rubbers occurs around a hydrostatic pressure of 2 MPa for any stress field. Within the styren butadiene rubber (SBR), cavitation appears at a hydrostatic pressure of 7–8 MPa.

Figure 4.17 presents the cavitation phenomenon.

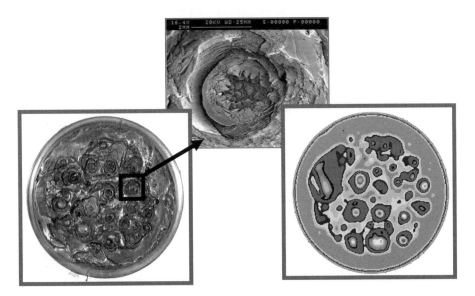

Figure 4.17. *Fatigue cavitation due to the hydrostatic pressure observed by scanning electron microscope and tomography [BAT 02]*

4.11. Conclusion

The fatigue damage of polymers and rubbers, which have only been used in the mechanical industry for 50 years, is much less well understood than metals. Polymers and rubbers do not react in the same way to a stress field. Increasing the average stress of a cyclic loading applied to a nature rubber increases its fatigue life. This is the opposite of what occurs in a metal! This chapter intended to show that neither the von Mises stress nor Griffith dissipated energy provide satisfying solutions for the prediction of fatigue crack initiation and propagation within these macromolecular materials. Some explanation is given from a phenomenological point of view.

4.12. Bibliography

[ARA 72] S. ARAD, J.C. RADON, L.E. CULVER, "Growth of the fatigue cracks in polycarbonate", *Polym. Eng. Sci.*, vol. 12, no. 3, pp. 193-198, 1972.

[BAT 02] C. BATHIAS, K. LE GORJU, "Fatigue initiation and propagation in natural and synthetic rubbers", *International Journal of Fatigue*, vol. 24, pp. 85-92, 2002.

[CHU 85] A. CHUDNOVSKY, A. MOET, "Thermodynamics of translational crack layer propagation", *J. Mater. Sci.*, vol. 20, pp. 630, 1985.

[DIL 50] J.H. DILLON, "Fatigue phenomena in high polymers", *Ado. Calloid Science*, vol. 3, pp. 219, 1950.

[GEN 92] A.N. GENT, *Engineering with Rubber*, Hanser Publishers, New York, 1992.

[HER 80] R.N. HERZTBERG, J.A. MANSON, *Fatigue of Engineering Plastics*, Academic Press, London, 1980.

[LAK 83] C.J. LAKE, "Aspects of fatigue and fracture of rubbers", *Prog. Rubber Technol*, vol. 45, pp. 89, 1983.

[LIN 72] M.J. LINDLEY, "Energy for crack growth in model rubber components", *Strain Analysis*, vol. 7, pp. 132, 1972.

[MUK 71] B. MUKHERGEE, D.J. BURNS, *Journal Exp. Mich.*, vol. II, pp. 433, 1971.

[RIV 53] R.S. RIVLIN, A.G. THOMAS, "Rupture of rubber", *J. Polymer Sci.*, vol. 10, pp. 291, 1953.

[ROS 64] B. ROSEN, *Fracture Processes in Polymeric Solids*, John Wiley and Sons, New York, 1964.

[SCH 95] R. SCHIRRER, *Les craquelures dans les polymères*, INPL, 395, 1995.

[STE 83] A. STEVENSON, "Fatigue crack growth in high load capacity laminates", *Int. J. Fracture*, vol. 23, pp. 47, 1983.

Chapter 5

Probabilistic Design of Structures Submitted to Fatigue

5.1. Introduction

The fatigue behavior of materials submitted to cyclic loading is a random phenomenon at any scale of description. When considering materials without microscopic defects, the fatigue initiation sites usually correspond to the creation of slip bands within the superficial grains and are influenced by the size and location of the grains as well as the roughness of the material surface. When considering materials with inclusions (e.g. carbides within nickel-based alloys) or defects (micro-cavities in cast steels), the latter may become preferential sites for crack initiation. The location and the size of these defects within an elementary material volume are naturally random ([BAT 10], Chapter 3).

Once the crack has been initiated at the microscopic scale, its transgranular propagation in stage I is controlled by the crystalline orientation of the grain and its neighbors: the fine description of this phenomenon can be deterministically modeled at the grain scale. Nevertheless, when considering the crack propagation at the macroscopic scale of the structure, the phenomenon is still random.

At the specimen scale, the different mechanisms that are presented above, cannot be predicted deterministically: the lifetime of a specimen of a particular material (i.e. whose composition is perfectly controlled) subjected to an identical loading, varies from one specimen to another: this is the scattering observed by any

---

Chapter written by Bruno SUDRET.

experimenter, and which can be represented by a scatter plot within the $(\log N,\ S)$ plane, where S is the loading amplitude and N the number of cycles to failure measured according to the standards, e.g. [AFN 90].

If the fatigue phenomenon is considered through the propagation of a crack under a cyclic loading at the macroscopic scale, it is usually observed that the parameters controlling the propagation (e.g. Paris law's parameters [PAR 63]) are random, as shown, for instance in [VIR 78] for aluminum.

Thus the random aspect of the fatigue phenomenon seems to be occurring at all scales of description. However, if we pay a close attention to the regulations in force regarding the behavior of loaded structures under fatigue conditions, they are based on *deterministic rules* (RCC-M code for the nuclear industry [AFC 00], AC25.571-1 rules of the *Federal Aviation Administration* for the space industry, etc.). Relying on so-called conservative criteria (which means that the design performed according to these criteria shall ensure safety by construction), they are based on simplified engineering models. In addition, academic work on fatigue often tends to explain deterministically how the fatigue phenomena occur.

From these initial observations, it seems important to develop a coherent probabilistic fatigue phenomenon approach based on a fine description of the physics at the desired scale and on a rigorous treatment of uncertainty. For the last ten years, the uncertainty treatment in physical models has been studied in various domains (probabilistic engineering mechanics, structural reliability, stochastic finite elements). It is therefore suitable to apply the general uncertainty quantification methods to the fatigue problems of materials and structures: this is what this chapter focuses on.

A general framework dealing with the treatment of the uncertainties in mechanical models is proposed in section 5.2. Different statistical treatment methods of fatigue data are then described in section 5.3. A probabilistic design methodology is then proposed in section 5.4. The uncertainty issue regarding the propagation of existing cracks is addressed in section 5.5, by considering some non-destructive test data to update the predictions.

5.2. Treatment of hazard in mechanical models

5.2.1. *General scheme*

The treatment of uncertainties in mechanical models has been studied for the past 30 years and has taken different names such as *random mechanics* [KRE 83], *structural reliability* [DIT 96, LEM 05, MEL 99], *stochastic finite elements* [GHA 91] or *sensitivity analysis* [SAL 00, SAL 04] depending on the research area. Under these different names, we can find a common methodology which allows us to come up with Figure 5.1 [SUD 07].

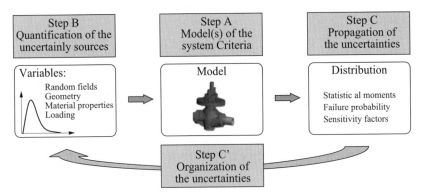

Figure 5.1. *Uncertainty treatment in mechanical models*

In the first step (called A), we usually define the *model* of the considered mechanical system, and especially its input parameters and its response (also named *quantity of interest*). We shall call x the vector of the input parameters, which may describe the geometry of the system (length of the elements, shape of the cross-sections, etc.), the behavior (elasticity moduli, parameters of the constitutive laws) and the applied loading. The response $y = \mathcal{M}(x)$ is usually a vector gathering displacements, stress and strain components, internal variables (strain hardening, damage indicators), etc. It may also contain some post-processed quantities (e.g. amplitudes of extracted cycles by the Rainflow method, cumulative damage, etc.). Criteria related to these quantities of interest are also usually defined (e.g. acceptable threshold for a specific response in the context of reliability analysis).

5.2.2. *Probabilistic model of the input parameters*

As the model and its input parameters have been defined, we shall now focus on those parameters that are uncertain and define a *probabilistic modeling*. The sources of uncertainty, which can be multiple, are usually split into two categories:

– The *epistemic* uncertainties, which are due to a lack of knowledge (e.g. lack of precision of the measurements, lack of data which may lead to insufficient reliable statistical sampling, etc.). They can usually be *reduced* in a sense since an acquisition of additional data or more precise data allows us to reduce them.

– The *random* uncertainties, which are intrinsic with respect to the observed parameter, and cannot be reduced. This is typically the case for the quantity "number of cycles to failure of a material under cyclic loading with a given amplitude": the more specimens are tested in the same experimental conditions, the more likely it is to find some extreme values of this number of cycles (low or high). This random uncertainty can also show a spatial variability as is the case for the properties of geomaterials.

Building a probabilistic model of the parameters (step B) consists of defining the probability distribution function (PDF) of the random vector X of the input parameters. When we do not have enough data to model the variability of a parameter, we can rely on *expert judgment*: we assume a certain shape of the distribution of the considered parameter (e.g. Gaussian distribution, uniform, lognormal, Weibull, etc.) and then we set up the mean value and the standard deviation of the distribution. To do so, guidelines are available in [JOI 02].

When data is available, we may use common *statistical inference* techniques [SAP 06]. In general, we search for the best distribution among selected families (for instance, using the maximum likelihood method) and then carry out goodness-of-fit tests to validate or discard the different choices. For fatigue issues, this type of statistical analysis is performed to process test data in order to establish the probabilistic Wöhler curves, as we will see in section 5.3. When the available sample set has a small size, we may combine the data with some prior information on the distribution (that represents *expert judgment*) by using *Bayesian statistics* [ROB 92, DRO 04].

Sometimes, the variability of the parameters of interest cannot be directly measured (e.g. parameters of the model of crack initiation), but this variability can be apprehended through the measurements of quantities of interest which depend on it (e.g. number of cycles to failure). In these cases we have to rely on *probabilistic inverse methods*. Broadly speaking these inverse problems are unsolved as yet. Some recent studies have recently been published in the field of fatigue though [PER 08, SCH 07]. Whatever the selected approach, step B eventually yields a probabilistic description of the input data as the probability density function $f_X(x)$ of the random vector of the input parameters X (see Appendix A for the elementary notions of probability theory).

5.2.3. *Uncertainty propagation methods*

Once both the mechanical model and the probabilistic model of its input parameters have been established, we shall focus on the *propagation of uncertainty* (step C in Figure 5.1). We are now ready to characterize the random response of the model, written as $Y = \mathcal{M}(\boldsymbol{X})^1$.

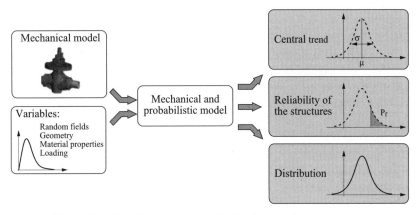

Figure 5.2. *Classification of the methods of uncertainty propagation*

Depending on what information we are looking for on Y (which is assumed to be a scalar quantity in this chapter for the sake of simplicity), we usually distinguish different types of analyses (Figure 5.2):

    – *Analysis of the central tendency*, where we mainly focus on the mean value μ_Y and on the variance σ_Y^2 of the quantity of interest Y (higher order statistical moments may also be computed). If the distribution of the response Y was Gaussian, these two scalars would completely characterize it. However, this assumption is usually *wrong* in practice: thus the statistical moments shall be used for what they are, without deducing any information on the distribution of Y from their values.

    – *Reliability analysis*, where we focus on the probability that the response is above a certain threshold $\bar{y}$. We then have to estimate the tail of the distribution of Y in this case. The associated probability (called failure probability) is usually low, with values ranging from 10^{-2} to 10^{-8}.

    – *Distribution analysis*, where we are looking to characterize the probability density function of Y completely.

Specific methods have been proposed for every type of problem, which are given below. For more information, see [SUD 07]:

    – The *Monte Carlo method* is the best known method of uncertainty propagation [RUB 81]. It allows us to resolve the problems presented above at least theoretically.

---

1. If the variability of some parameters of the model is equal to zero or is negligible, we can consider them as deterministic and gather them within a vector $\boldsymbol{d}$, and we shall write $Y = \mathcal{M}(\boldsymbol{X}, \boldsymbol{d})$.

It relies on the simulation of random numbers by specific algorithms which generate a sample set of input parameters conforming to the probabilistic model built in step B (i.e. according to the distribution f_X). In general, this method is quite efficient for the calculation of the first statistical moments (mean value and variance) but becomes computationally expensive when considering reliability or distribution analysis.

– The *perturbation method*, which was developed in mechanics in the 1980s, relies on a Taylor series expansion of model $\mathcal{M}$ around the nominal value of parameters x_0 and allows the mean value and standard deviation of the quantities of interest to be efficiently estimated. The gradients of model $\mathcal{M}$ with respect to the input parameters need to be calculated. This method is nowadays used efficiently when these gradients are directly implemented into the finite element codes used in industrial applications (e.g. Code_Aster [EDF 06]).

– The resolution of reliability problems, i.e. the estimation of the probability of failure historically led to some specific methods back in the 1970s. The FORM/SORM methods (respectively *first* and *second order reliability method*) were established in the 1980s and have been used by some industries (offshore, nuclear industry) for about 15 years. They are *approximation* methods of the distribution tail of the response and do not allow us to estimate the quality of the obtained result. They are usually coupled to some advanced simulation methods (directional simulation, importance sampling, subset simulation, etc.) [LEM 09].

– Initially introduced in the *spectral stochastic finite elements* method [GHA 91], representations of response Y by polynomial chaos expansions are nowadays a promising approach for the treatment of uncertainties. The principle is to consider the random response Y within a suitable functional space of random variables, in which a basis is built. Response Y is then completely represented through its "coordinates" in this basis, which are the coefficients of the polynominal chaos expansion. These coefficients may be calculated by some non-intrusive methods from a limited number of model evaluations, namely $\mathcal{Y} = \{y^{(i)} = \mathcal{M}(x^{(i)}), i = 1, \dots, N\}$. The analytical post-processing of the coefficients provides the distribution of Y at no additional calculation costs as well as its statistical moments, the probabilities of exceeding a threshold, etc. [SUD 07].

– Most of the methods of uncertainty propagation give, as by-products (with almost no additional calculation), some information on the relative significance of the input parameters of the model: we call these by-products *importance factor* or *sensitivity indices* depending on the methods. This hierarchization (or ranking) of the parameters according to their importance is called "step C'" in Figure 5.1.

5.2.4. *Conclusion*

In this introductory part, we tried to focus on defining a general framework for dealing with uncertainties in mechanical models. We defined the different necessary ingredients and then presented the most commonly used calculation methods.

The general scheme will now be applied to different problems related to the fatigue of materials and structures.

5.3. Plotting probabilistic S–N curves

5.3.1. *Introduction*

The experimental points obtained from fatigue testing are usually plotted in the "amplitude S/number of cycles to failure N" space in order to get the so-called Wöhler curves ([BAT 10], Chapter 2). In France, the statistical treatment obeys a standard [AFN 91] (mainly inspired by the work of Bastenaire [BAS 60]), which provides methods to obtain the endurance limit using the staircase method (Section 5 of the standard), the median Wöhler curve (Section 6), or some probabilistic $S - N$ curves (ESOPE method, Section 9).

The latter is summarized below in section 5.3.2. An alternative approach for establishing probabilistic $S - N$ curves is proposed. This approach is based on the studies performed independently by Guédé [GUÉ 05], Perrin [PER 08], and Pascual and Meeker [PAS 97, PAS 99]. It has recently been used in industry by EDF and EADS [SCH 06, SCH 07].

Regardless of the approach, we consider a sample set of measurements:

$$\mathcal{E} = \{(S_i, N_i), i = 1, \ldots, Q\} \tag{5.1}$$

where N_i is the number of cycles to failure measured under alternate stress of amplitude S_i and Q is the size of the sample set. We consider that the sample set contains test results carried out in the same experimental conditions, allowing us to treat all the experimental points as a whole. We will also consider the data obtained in the case when the fatigue test has been stopped before failure has occurred. This so-called censored data is denoted by N_i^*.

5.3.2. *ESOPE method*

The ESOPE method, recommended by the AFNOR standard (A 03-405, Section 8) [AFN 91] is based on the following assumption: within a sample $\mathcal{E}$ as presented above, we *suppose* that the fraction $F(S, N)$ of the specimens, which failed *before* N cycles of amplitude S, has the following particular form:

$$F(S, N) = \Phi\left(\frac{S - \mu(N)}{\sigma}\right) \tag{5.2}$$

where σ is a distribution parameter, Φ is the standard normal (Gaussian) cumulative distribution function (see equation [5.31]) and $\mu(N)$ is a curve whose shape is prescribed by the analyst and whose parameters are estimated from $\mathcal{E}$. We should note that the curve representing the median lifetime $N_{50\%}$ (corresponding to points such that $F(S, N) = 0.5$) corresponds to $S = \mu(N)$, according to equation [5.2], since $\Phi^{-1}(0.5) = 0$. The median curve gives, for each stress level S, the value $N_{50\%}$ such that there is as much chance of a specimen under loading S failing before or after $N_{50\%}$ cycles.

On the other hand, the quantity $F(S, N)$ is an estimation of the cumulative distribution function $F_{N_S}(N; S)$ of the random variable $N_S(\omega)$ defined[2] as the number of cycles to failure of the considered material, under loading amplitude S. As a consequence, according to [5.2], the isoprobabilistic failure curves $N_p(S)$ defined as

$$\mathbb{P}\left(N_S(\omega) \leq N_p(S)\right) = p \qquad [5.3]$$

are also defined within the plane (N, S) by the following equation:

$$S = \mu(N) + \sigma \, \Phi^{-1}(p) \qquad [5.4]$$

It is clear that they may be obtained by vertically translating the median curve by a quantity $\sigma \, \Phi^{-1}(p)$. In practice, we usually take a parametric form for this median lifetime, e.g. as proposed by the AFNOR 03-405 norm:

$$\mu(N) = a + b \, N^c \qquad [5.5]$$

By combining equations [5.2] and [5.5] we get:

$$F_{N_S}(N; S) = \Phi\left(\frac{S - [a + b \, N^c]}{\sigma}\right) \qquad [5.6]$$

We then estimate parameters a, b, c using the *maximum likelihood* method 5 (see section 5.7.3) from which the probability density function $f_{N_S}(N; S)$ can be derived. We can then write the likelihood of parameters a, b, c and estimate these parameters by maximizing the likelihood function.

NOTE: Equation [5.6] is often given in the following interpretation: "the variability of stress S, for a fixed lifetime N is normally distributed with a standard value σ." Actually, this assertion does not make any sense since, from an experimental point of view, the loading amplitude S is fixed (and not at all random!). Conversely, lifetime N under loading S is a random variable since it is related to the occurrence of cracking mechanisms at the microscopic scale that cannot be deterministically described at the macroscopic scale, as explained in the introduction.

---

2. Throughout the whole chapter, the notation ω highlights the random property of the considered quantity. When there is no ambiguity, ω can be omitted for the sake of simplicity.

5.3.3. *Guédé-Perrin-Pascual-Meeker method (GPPM)*

The ESOPE method, which was described above, does not explicitly provide the probability density function of the random variables $N_S(\omega)$ modeling the lifetime of the specimens. Thus it is not suitable for a complete probabilistic assessment of a structure under fatigue loading, as we will see later on. The problem is better addressed by the direct modeling of the physically uncertain quantities, namely the number of cycles to failure of the specimens under loading amplitude S, i.e. by the inference of the random variables $N_S(\omega)$.

Several formulations were proposed in [SUD 03a, GUÉ 05] and were further elaborated by Perrin [PER 05]. The final formalism is actually rather close to the one proposed independently by Pascual and Meeker [PAS 97, PAS 99], which explains the GPPM acronym.

5.3.3.1. *Guédé's assumptions*

The approach by Guédé is split into different steps:

– The *choice of the distribution* for the variables $N_S(\omega)$: a lognormal distribution is usually selected (the logarithm of the number of cycles to failure is supposed to follow a Gaussian distribution). We should mention that other choices are possible (including the Weibull law). The hypothesis has to be validated *a posteriori* by performing some statistical tests.

– The *description of the parameters* of the distribution of $N_S(\omega)$ as a function of S. The initial work carried out by Sudret *et al.* [SUD 03a] assumes that the mean value $\lambda(S)$ of $\ln N_S$ can be written as follows (Stromeyer's formula):

$$\lambda(S) = A \ln(S - S_D) + B \qquad [5.7]$$

where constants A, B and S_D are to be determined from the data. We may then consider either a constant standard deviation [LOR 05] (which is reasonable if the sample set contains mainly points in the low cycle domain), or a variable standard deviation which is a function of the amplitude of the loading S:

$$\sigma(S) = \delta \, \lambda(S) \qquad [5.8]$$

This last equation allows us to represent the large scattering of the data usually observed in the high cycle fatigue domain.

– Regarding the type of dependency between the random variables $N_S(\omega)$ for different values of S, the hypothesis of *perfect dependency* has been adopted: it corresponds to the intuitive idea that if we could test the same specimen at different loading levels (which cannot actually be performed since fatigue tests are destructive), the fatigue strength would be uniformly good or bad, meaning that the number of cycles to failure would be, regardless of S, away from the median value with more or less the same proportions.

5.3.3.2. *Model identification*

The previous assumptions allow us to write the random variable $N_S(\omega)$ as follows:

$$\ln N_S(\omega) = \lambda(S) + \sigma(S)\,\xi(\omega) = (A\,\ln(S - S_D) + B)(1 + \delta\,\xi(\omega)) \qquad [5.9]$$

where $\xi(\omega)$ is a standard Gaussian variable (i.e. with a mean value equal to zero and a unit standard deviation). At this stage, the probabilistic model depends on the 4 parameters A, B (shape of the median curve), S_D (asymptote, which is considered as an infinite endurance limit) and δ (coefficient of variation of $\ln(N_s)$). Having a single variable $\xi(\omega)$ in [5.9] (and not one variable $\xi_S(\omega)$ for each amplitude) corresponds to the perfect dependency assumption described previously. From equations [5.7]–[5.9], we can get the probability density function of variable $N_S(\omega)$:

$$f_{N_S}(n, S; A, B, S_D, \delta) = \frac{1}{\delta\,[A\,\ln(S - S_D) + B]\,n}\,\varphi$$
$$\varphi\left(\frac{\ln n - [A\,\ln(S - S_D) + B]}{\delta\,[A\,\ln(S - S_D) + B]}\right) \qquad [5.10]$$

where $\varphi(x) = e^{-x^2/2}/\sqrt{2\pi}$ is the standard normal probability density function.

For a given sample set $\mathcal{E}$ (see equation [5.1]), the *likelihood* of parameters A, B, S_D and δ reads:

$$\mathsf{L}(A, B, S_D, \delta; \mathcal{E}) = \prod_{i=1}^{Q} f_{N_S}(N_i, S_i; A, B, S_D, \delta) \qquad [5.11]$$

The maximum likelihood method consists of estimating the unknown parameters A, B, S_D, δ by *maximizing* the previous quantity (or by minimizing the log-likelihood $-2\ln\mathsf{L}$). The intuitive interpretation of the method is simple: it leads to the choice of the parameters which maximize the probability of having observed the available sample set $\mathcal{E}$. It is worth noting that the censored data N_i^* (no failure observed before N_i cycles with an amplitude S_i) can be used within equation [5.11] by replacing the probability density $f_{N_S}(N_i, S_i; A, B, S_D, \delta)$ by $1 - F_{N_S}(N_i^*, S_i; A, B, S_D, \delta)$, where the cumulative distribution function F_{N_S} is given by:

$$F_{N_S}(n, S; A, B, S_D, \delta) = \Phi\left(\frac{\ln n - [A\,\ln(S - S_D) + B]}{\delta\,[A\,\ln(S - S_D) + B]}\right) \qquad [5.12]$$

and $\Phi(x)$ is the standard normal cumulative distribution function.

Once the parameters have been estimated from the sample set (they are denoted with a hat from now on), the $S - N$ curves can be naturally obtained from equation [5.9]. The equation of the iso-probability failure curve p is

$$N_p(S) = (\hat{A}\ln(S - \hat{S}_D) + \hat{B})(1 + \hat{\delta}\,\xi_p) \qquad [5.13]$$

where $\xi_p = \Phi^{-1}(p)$ is the quantile of level p of the standard normal distribution. The median curve ($\xi_p = 0$) and, for instance, the quantiles at 5% and 95% ($\xi_p = \pm 1.645$) can then be easily plotted.

5.3.3.3. GPPM method

In the previous section parameter S_D was introduced as a fitting parameter in order to characterize an asymptotic behavior in the high cycle domain. However, this parameter could be considered from two different perspectives:

– It may be viewed as a *deterministic* fitting parameter, i.e. a stress amplitude such that a fatigue test carried out below this level would never lead to failure. This is what was assumed in the previous section. Then the iso-probability curves have the same horizontal asymptote at S_D in this case.

– The fatigue limit may also be considered as a "true" material parameter, whose value differs from one specimen to another: it would be the critical amplitude such that, *for the considered specimen*, failure never occurs for any cyclic loading with an amplitude lower than this value. In that case, S_D should be considered a random variable whose realization is different (and obviously unknown) for every tested specimen. In this context it is meaningful to try to infer the distribution of S_D from data together with the other parameters describing the probabilistic Wöhler curves: this is the goal of the GPPM model.

The same assumptions as in section 5.3.3.1 are taken into account except for parameter S_D which is now modeled as a random variable, with a probability density function $f_{S_D}(s_D; \boldsymbol{\theta})$, where $\boldsymbol{\theta}$ is the vector of the distribution parameters (e.g. $\boldsymbol{\theta} = (\lambda_{S_D}, \zeta_{S_D})$ for a lognormal law). The probability density function given in equation [5.10] becomes *conditional* to these parameters, and is written $f_{N_S|S_D}(n, s_D, S; A, B, \delta)$. In order to calculate the likelihood (equation [5.11]), the unconditional distribution f_{N_S} has to be used; it can be obtained by performing the following integration:

$$f_{N_S}(n, S; A, B, \delta, \boldsymbol{\theta}) = \int f_{N_S|S_D}(n, s_D, S; A, B, \delta)\, f_{S_D}(s_D; \boldsymbol{\theta})\, ds_D \qquad [5.14]$$

In the end, the solution of the maximum likelihood problem provides both the estimators of the parameters controlling the failure iso-probability curves $\hat{A}, \hat{B}, \hat{\delta}$ and those of the distribution of the endurance limit S_D (see details in [PER 08]).

5.3.4. *Validation of the assumptions*

In both the ESOPE method and the GPPM approach, some hypotheses are made regarding the shape of the distributions. The theory of the goodness-of-fit tests [SAP 06, Chapter 14] allows these hypotheses to be validated or rejected. These hypotheses (usually called "null hypothesis H_0") are as follows:

ESOPE H_0: variable $S - [a + b N^c]$ follows a normal distribution with a standard value σ (see equations [5.2] and [5.5]). For every point of the sample $\{(N_i, S_i), i = 1, \ldots, Q\}$, the previous quantity is calculated and a sample is then obtained $\{\xi_i, i = 1, \ldots, Q\}$ where H_0 is tested.

Guédé H_0: variable $\xi(\omega) \equiv ((\ln N_S(\omega)/[A \ln(S - S_D) + B]) - 1)/\delta$ follows a standard normal distribution. For every point of the sample $\{(N_i, S_i), i = 1, \ldots, Q\}$, the previous quantity is calculated and a sample is then obtained $\{\xi_i, i = 1, \ldots, Q\}$ whose normality is tested.

GPPM H_0: for each amplitude S, the random variable $N_S(\omega)$ follows the distribution given by equation [5.14].

It is then possible to quantitatively compare the different methods of the various assumptions and to validate *a posteriori*.

5.3.5. *Application example*

To illustrate the different methods presented above, the results from Perrin [PER 08] are reported in this section. A sample set of 153 test results on austenitic steel specimens at 20°C was presented. As presented in Figure 5.3, the different approaches give close results in terms of median curve.

The curves of the quantiles at 2.5% and 97.5% are given in Figure 5.4. A significant difference can be observed regarding the general shape of the associated confidence intervals. Indeed the ESOPE method seems to underestimate the variability in the low cycle domain.

For different amplitudes S, the probability density function of the lifetime $N_S(\omega)$ can be plotted using different methods [PER 08]. In particular, the goodness-of-fit tests lead to the rejection of the ESOPE model, which is not the case with the two other models. The median and quantile curves obtained using both the Guédé and the GPPM methods are very similar, which is the same regarding the probability density

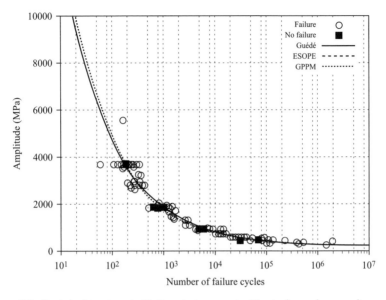

Figure 5.3. *Statistical treatment of fatigue tests on austenitic steel samples – median curves obtained with the ESOPE method and the Guédé and GPPM methods*

functions of the lifetime for some high amplitude levels. The difference between these approaches appears for the amplitudes close to the endurance limit. The latter is estimated to be 236.9 MPa by the Guédé method and by a lognormal variable of mean value 257.8 MPa and standard value of 28.4 MPa by the GPPM method.

5.3.6. *Conclusion*

The statistical treatment of test data is a step that cannot be ignored in the design of structures with respect to fatigue. In this section, the principles of the ESOPE method, which is the most commonly used in the industrial domain, were presented. An alternative formulation (named GPPM), based on the direct inference of the distributions of the lifetimes $N_S(\omega)$ was proposed and allows a random endurance limit to be considered, whose parameters are estimated jointly with those describing the median Wöhler curve. This approach leads to the treatment of an experimental scatter plot (including censored data) without having to rely on specific amplitude levels and many points for each level. This is not the case for the ESOPE method which is based on the estimation of the fraction of broken specimens before N cycles, for each amplitude level S and thus requires a large amount of data at each level.

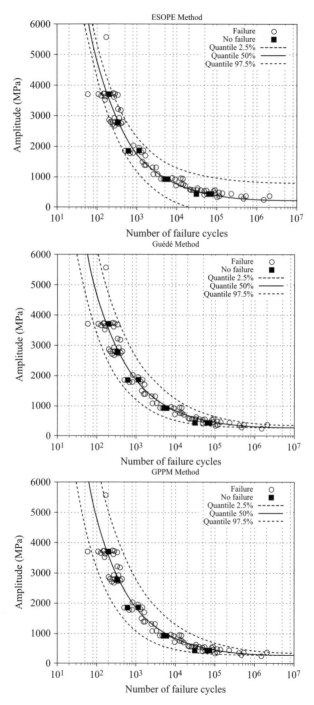

Figure 5.4. *Statistical treatment of fatigue tests on austenitic steel samples –
quantile curves at 2.5% and 97.5% for the different methods*

5.4. Probabilistic design with respect to crack initiation

5.4.1. *Introduction*

Defining Wöhler curves from the fatigue test data is usually the first step towards the design of structures subjected to fatigue. Broadly speaking, in order to design a structure, the applied loading (periodic or random loading) and its effects (stress analysis, extraction of the stress cycles) have to be characterized, and then the damage due to these stress cycles has to be calculated. In the case of a probabilistic analysis, the same procedure will be carried out including the sources of uncertainty at each step.

To do so, the *deterministic* design of fatigue of a component of a nuclear plant (RCC-M code [AFC 00]) and its transposition to a probabilistic equivalent (Guédé's work [GUÉ 05, SUD 05]) is taken as an example. According to the general scheme for managing uncertainty presented in section 5.2.1, the deterministic model (step A) will be described first, then the different sources of uncertainties will be characterized (step B) and then the uncertainties will be propagated through the model to evaluate the reliability of the structure subjected to fatigue.

It is important to note that the whole methodology does not depend on the models used at each step of the deterministic calculation, which are much simplified here for the sake of clarity. The same general framework has been recently applied by Schwob [SCH 06, SCH 07] using some multiaxial fatigue criteria in collaboration with EADS and by Perrin [PER 06] in collaboration with Renault. It is also worth noting that the method called *stress – resistance* [THO 99], which was developed for the automotive industry by PSA Peugeot Citroën, applies the same concepts, just in a simplified manner.

5.4.2. *Deterministic model*

The main steps regarding the official standards for designing pipes subjected to fatigue are listed below [AFC 00]:

Description of the loading: the main issue here is the thermal fatigue induced by temperature fluctuations within the pipes. It is necessary to determine the temperature history at the inner walls of the pipes for each particular operating sequence. This temperature history can be obtained by computational fluid dynamics (CFD) or using some measurements on extrapolated scale models, etc.

Mechanical model: from the description of the structure's geometry (e.g. internal radius, thickness), of some material properties (Young's modulus, Poisson ratio, parameters of the elasto-plastic constitutive laws), of the boundary conditions (fluid/structure heat transfer coefficient) and of the loading, the strain and stress fields are calculated as a function of time.

Extraction of fatigue cycles: from the stress tensor the equivalent stress history is obtained by means of the Tresca criterion. Then the cycles are extracted using the rainflow method [AMZ 94], yielding a sequence of amplitudes S_i, $i = 1, \ldots, N_c$. The values obtained are then corrected in order to consider the effect of the mean stress (Goodman line within Haigh diagram).

Choice of the design curve: in the RCC-M standard the design curve $N_d(S)$ can be obtained by modifying the median Wöhler curve $N_{bf}(S)$ obtained from some tests on specimens using *coefficients* that are called *passage factors*, which empirically take into account all the factors leading to a reduced lifetime for the structure compared to the specimens, and aiming to be conservative. The RCC-M code then defines the design curve with the following equation:

$$N_d(S) = \min(N_{bf}(S)/\gamma_N \ , \ N_{bf}(\gamma^S S)) \tag{5.15}$$

The value of the passage factors are respectively $\gamma^N = 20$ (reduction of the number of cycles in the low cycle domain) and $\gamma^S = 2$ (increase of the applied stresses in the high cycle domain).

Cumulative damage rule: Miner's linear cumulative damage rule is applied [MIN 45]: each cycle with an amplitude S_i is supposed to generate the elementary damage $d_i = 1/N_d(S_i)$, and this damage is supposed to accumulate at every cycle extracted using the rainflow method:

$$D = \sum_{i=1}^{N_c} d_i = \sum_{i=1}^{N_c} 1/N_d(S_i) \tag{5.16}$$

In order to apply the design criterion, the accumulated damage has to remain lower than 1. When the loading applied to the structure is made of some sequences which are supposed to be identical, the cumulative damage D_{seq} can then be calculated for each single sequence. Then the lifetime of the structure can be cast as $T_d = 1/D_{seq}$, corresponding to an acceptable number of sequences.

The different steps of the calculation are presented in Figure 5.5.

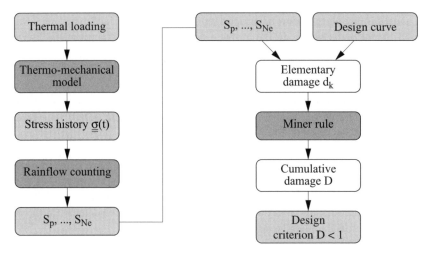

Figure 5.5. *Simplified deterministic scheme for the design of a pipe submitted to thermal fatigue*

5.4.3. *Probabilistic modeling of uncertainties*

The computational flowchart presented in the above section involves many parameters which are not well known in practice (e.g. the fluid/structure heat transfer coefficient) or are intrinsically random (e.g. the lifetime of the fatigue test specimens). It is then necessary to classify all the sources of uncertainty, and then to model them within a consistent probabilistic frame depending on the available data (Step B of the general scheme, Figure 5.1). By analyzing the calculation process step by step, the following parameters will be successively presented:

Thermal loading: to design some part of a circuit for a specific operating range, a thermal dimensioning signal has to be defined, either from some measurements on real structures or scale models, or from some operating scenarios or related thermohydraulic simulations. The uncertainty can be introduced by modeling the fluid temperature by a *random process*. To do so, the properties of the process can be identified from some long enough measurements (either within the time domain or within the frequency domain through the power *spectral density*). The problem can also be simplified if a sufficiently long deterministic temperature trajectory correctly representing the thermal fluctuations can be determined.

Mechanical model: the parameters which are usually involved in the definition of the mechanical model (geometric dimensions, elastic or elasto-plastic properties of the materials) present a variability which can be modeled using random variables. The uncertainty in the geometry is related to the pipe manufacturing tolerances.

The uncertainties of the material properties can be characterized by some tests on the considered materials or, in the absence of available data, found in the literature [JOI 02].

Design curve: The design curve was defined earlier from the specimen median Wöhler curve from the passage factors. In section 5.3, a meticulous statistical analysis methodology yielding the probabilistic Wöhler curves was presented. The passage factors shall now be studied.

Passage factors $\gamma^N = 20$ and $\gamma^S = 2$, defined above, have been introduced to the deterministic design in order to conservatively cover two effects of a very different nature: the natural variability of the lifetime under fatigue between laboratory specimens on the one hand and the structure within its real environment on the other hand. These factors can then be split as follows:

$$\gamma^N = \gamma^N_{disp} \cdot \gamma^N_{passage}$$
$$\gamma^S = \gamma^S_{disp} \cdot \gamma^S_{passage} \qquad\qquad [5.17]$$

In these equations, γ_{disp} and $\gamma_{passage}$ correspond to the part connected to each effect. Some empirical decompositions can be found in the literature, for instance $\gamma^N_{disp} = 2$ when $\gamma^N = 20$ and $\gamma^S_{disp} = 1.19$ when $\gamma^S = 2$ [COL 98]. However, there is no real agreement regarding this topic as the data are mostly empirical and much connected to the type of material tested. The additional variables $\gamma^S_{passage}$ and $\gamma^N_{passage}$ are called specimen-to-structure passage factors and allow us to consider the effects of the size of the structure, its surface finish and the environmental conditions (especially regarding temperature and chemistry in nuclear engineering, etc.).

It is obvious that these passage factors that connect the crack initiation time of a specimen to that of a structure made of the *same* material cannot be physically measured. In order to get a rigorous probabilistic representation, it is then necessary to rely on some probabilistic inverse methods, which allow the factors and their distributions to be estimated from both data on specimens and structures (e.g. some pipe scale models that are typical of the considered real system, like the INTHERPOL tests described in [CUR 04, CUR 05]). Details about this identification technique are not reported in this chapter, see [PER 07a] for a detailed presentation.

5.4.4. *Random cumulative damage*

The accumulated damage D can be considered as the result of the computational chain, as shown in Figure 5.5. If we now consider the input parameters of each sub-model of this chain to be random (thermomechanical model, extraction of the cycles, design curve, etc.), the accumulated damage becomes random. The natural definition of the random elementary damage related to a single cycle with a fixed amplitude S is:

$$d(S, \omega) \equiv 1/N^{struc}(S, \omega) = 1/\min \left[N_S(\omega)/\gamma^N_{passage} \, , \, N_{\gamma^N_{passage}} \cdot S(\omega) \right] \quad [5.18]$$

In this equation, $N^{struc}(S, \omega)$ is the lifetime of the structure under a loading with a constant amplitude S, which is connected to the probabilistic Wöhler curve $N_S(\omega)$ by the passage factors. If the thermomechanical calculation followed by the rainflow counting leads to a number of cycles N_c (possibly random), the accumulated random damage $D(\omega)$ is then written:

$$D(\omega) = \sum_{i=1}^{N_c} d(S_i(\omega), \omega) \quad [5.19]$$

In this equation, the random property of the damage comes from both:

– the randomness of the loading, which is propagated through the mechanical model for evaluating the amplitude of the cycles $\{S_k(\omega), \, k = 1, \, \ldots, \, N_c\}$;

– the randomness of the material fatigue strength, through the Wöhler curve.

In the case of a stationary random loading, for some loading sequences which are long enough, the number of extracted cycles N_c becomes high and can then be considered in a first approximation as being deterministic [TOV 01]. The amplitudes of the extracted cycles can then be continuously represented by their probability density function $f_S(s)$. This leads to the "continuous" definition of the random accumulated damage:

$$D(\omega) = \int_0^\infty \frac{N_c \, f_S(s) \, dS}{N^{struc}(S, \omega)} = N_c \, \mathrm{E}_S \left[\frac{1}{N^{struc}(S, \omega)} \right] \quad [5.20]$$

where $\mathrm{E}_S[.]$ stands for the mathematical expectation with respect to the probability density function of amplitudes $f_S(s)$. Assuming Miner's linear cumulative damage assumption and a significant number of independent extracted cycles, the cumulative damage defined by equation [5.19] has been proven to converge towards the one defined by equation [5.20] [SUD 03b].

The previous continuous formulation can be successfully applied to a fatigue assessment in the frequency domain. From the power spectral density (PSD) of the loading, and if the mechanical model is linear, the PSD of the resulting equivalent stress can be obtained. Some empirical formulae, such as those proposed by Dirlik [BEN 06, DIR 85], allow the probabilistic density of the extracted cycles using the rainflow method to be constructed, which can then be substitued into equation [5.20]. Readers are referred to the work by Guédé [GUÉ 05, Chapter 6, GUÉ 07].

5.4.5. Application: probabilistic design of a pipe under thermal fatigue conditions

5.4.5.1. Problem statement and deterministic model (step A)

To illustrate the different concepts presented in this section, a piece of pipe, that is of a circuit of a pressurized water reactor, will be considered. The results obtained by [GUÉ 05], Chapter 7, are reported. The reader can check this reference for a comprehensive parametric analysis of the problem under different loadings, as well as for a comparison of the probabilistic approaches within both time and frequency domains.

Let us consider a section of a pipe with a radius R_{int} and a thickness t, subjected to a fluid temperature at the inner wall modeled by a random Gaussian process $\theta(t, \omega)$, whose average temperature is equal to 130°C and standard deviation is equal to 20°C. This process is a pseudo-white noise whose power spectral density is constant within the [0 ; 5 Hz] interval. From these data, a realization $\theta(t)$ is simulated on a long enough time interval ([0 ; 360 s] here) and this signal is then considered as being periodically reproduced. The first 10 seconds of the signal are represented in Figure 5.6. The component service duration is considered to be equal to $N_{seq} = 10,000$ sequences of 360 seconds.

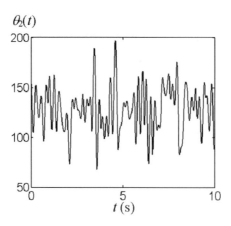

Figure 5.6. *Probabilistic design of a pipe submitted to thermal fatigue temperature history*

The stress state within the pipe can be calculated using a 1D axi-symmetrical model with generalized plane strains (the strain component ε_{zz} is assumed to be constant within the thickness). According to the RCC-M code, an elastic calculation is carried out. The fluid temperature is transferred to the inner wall of the pipe through a fluid/structure heat transfer coefficient. The external wall of the pipe is insulated (preventing the loss of heat). Due to the simplified form of the model, the obtained stress tensor is diagonal and its components are varying synchronously in time. The orthoradial stress time history $\sigma_{\theta\theta}(t)$ may be considered to perform the rainflow counting. The amplitudes of the extracted cycles are corrected by the Goodman line within the Haigh diagram in order to consider the average stress.

5.4.5.2. Probabilistic model (step B)

The different random variables modeling the uncertainties on the parameters of the model are presented in Table 5.1. The choice of the probability distribution functions and their parameters (step B) was performed as follows[3]:

– The properties of the materials are modeled by some lognormal laws which are usually well adapted to this type of parameter (being positive). The coefficients of variation have to be fixed by experts. The geometry parameters are also modeled by lognormal distributions, with a variation coefficient made high on purpose.

– The physically bounded variables (Poisson ratio, passage factors) are modeled by Bêta distributions.

– The probabilized Wöhler curve (equation [5.9]) was established by the Guédé method (section 5.3.3.1) and is written as follows:

$$N(S, \omega) = \exp\left[(-2.28 \log(S - 185.9) + 24.06)(1 + 0.09\,\xi\,(\omega))\right] \qquad [5.21]$$

where $N(S, \omega)$ is the crack initiation time of the specimens.

5.4.5.3. Reliability and sensitivity analysis (step C&C')

The probability of crack initiation is studied for a prescribed number of operating sequences N_{seq}. To do so, the following limit state function g is defined (the negative values of g correspond to the values of the parameters leading to failure):

$$g(N_{seq}, X) = 1 - N_{seq}\, d_{seq}(X) \qquad [5.22]$$

where $d_{seq}(X)$ stands for the random cumulative damage related to a sequence of 360 seconds of operation, and X stands for the vector of the random variables listed in Table 5.1.

By using the FORM method (Appendix A, section 5.7), a probability of crack initiation is obtained: $P_f = 8.59.10^{-2}$ for a service life of 10,000 sequences. This result is confirmed by some importance sampling, which finally give $P_f = 7.65.10^{-2}$.

The FORM method used for the calculation of the initiation probability also provides the *importance factors* of the different random variables, which allows the

---

3. The data used for the calculation (especially the distribution parameters) does not represent any real structure.

Parameter	Distribution	Mean value	V.C †
Internal radius R_{int}	lognormal	127.28 mm	5%
Thickness t	lognormal	9.27 mm	5%
Heat capacity ρC_p	lognormal	4,024,000 J/kg	10%
Thermal conductivity λ	lognormal	16.345 W.m^{-1}.K^{-1}	10%
Heat transfer coefficient H	lognormal	20,000 W.K^{-1}.m^{-2}	30%
Young's modulus E	lognormal	189,080 MPa	10%
Poisson rotio ν	Bêta [0.2; 0.4]	0.3	10%
Thermal dilatation coefficient α	lognormal	16.95 10^{-6}	10%
Yield stress S_y	lognormal	190 MPa	10%
Ultimate strength S_u	lognormal	496 Mpa	10%
Passage factor $\gamma_{passage}^N$	Bêta [7; 11]	9.39	10%
Passage factor $\gamma_{passage}^S$	Bêta [1; 2]	1.68	10%
Scattering of the fatigue data ξ (equation [5.9])	Gaussian	0	Standard value: 1

† coefficient of variation, equal to the ratio of the standard deviation and the mean value.

Table 5.1. *Probabilistic design of a pipe submitted to thermal fatigue: probabilistic model of the parameters*

Parameter	Importance factor (%)
Scattering of fatigue data ξ	40.3
Heat transfer coefficient H	20.8
Passage factor $\gamma_{passage}^S$	13.7
Young's modulus E	8.9
Thermal dilatation coefficient α	8.9
Other variables	7.4

Table 5.2. *Probabilistic design of a pipe submitted to thermal fatigue: importance factors*

input parameters of the model to be ranked according to their respective contribution to the fatigue. These normalized factors (given in percentages) are listed by decreasing order in Table 5.2.

Beyond the strict numerical values of these importance factors (which depend on the probability density functions chosen for modeling the input parameters), the orders of magnitude as well as the obtained classification lead to some observations. It clearly appears that the parameters that mainly explain the variability of the structure crack initiation time are the scattering of the endurance of the specimens (modeled in the probabilistic Wöhler curves). Then there are the fluid/structure transfer coefficient, the passage factor $\gamma_{passage}^S$ increasing the stress amplitudes in the high cycle domain, and finally the Young's modulus and the thermal dilatation coefficient. The other parameters have a negligible importance, which means that their variability does not contribute to the variability of the cumulative damage (and thus to the initiation probability). As a consequence, they can be considered to be deterministic in this type of analysis.

5.4.6. *Conclusion*

In this section, we propose a general framework to address the issue of uncertainties in the design of structures submitted to fatigue, stressing every uncertainty source observed in the computational chain. As an illustrative example, the probability of crack initiation within a pipe submitted to thermal fatigue was estimated. A quantification of the influence of every uncertain parameter on this probability was obtained. Even if caution is required for the values, some usefully qualitative conclusions can still be drawn for the comprehension of the problem.

5.5. Probabilistic propagation models

5.5.1. *Introduction*

As observed for crack initiation, the propagation of pre-existing cracks under cyclic loading shows some randomness. An extensive experimental study performed by Virkler *et al.* [VIR 78] on some alumina 2024-T3 specimens clearly shows the scattering of crack propagation among identical specimens (Figure 5.7): 68 pre-cracked (initial size $a_0 = 9$ mm) rectangular sheets with a length of $L = 558.8$ mm, a width of $w = 152.4$ mm and a thickness of $t = 2.54$ mm) are loaded under tension-compression ($\Delta\sigma = 48.28$ MPa, $R = 0.2$). The tests are stopped when the length of the crack reaches 49.8 mm. It can be observed that this final length is reached between 223,000 and 321,000 cycles depending on the specimens with a almost continuous distribution in between.

By observing Figure 5.7, two types of randomness can be noticed regarding crack propagation:

– a global scattering of the curves, whose shapes are similar although the number of cycles varies;

– for each curve, a local irregularity, which shows that the successive increments of the crack size along a given trajectory are also random.

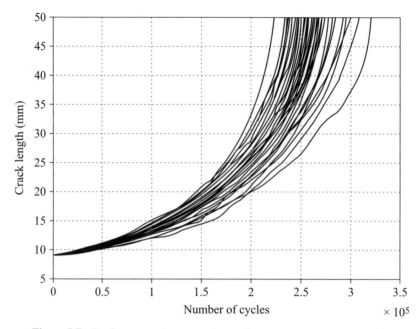

Figure 5.7. *Crack propagation – experimental propagation curves according to Virkler et al. [VIR 78]*

5.5.2. *Deterministic model*

Crack propagation under cyclic loading is usually modeled using the Paris-Erdogan law [PAR 63] ([BAT 10], Chapter 6):

$$\frac{da}{dN} = C\left(\Delta K\right)^m \qquad\qquad [5.23]$$

In this equation, a is the length of the crack, ΔK is the amplitude of the stress intensity factor for a cycle with an amplitude equal to $\Delta\sigma$ and (C, m) are the typical parameters of the studied material. In the case of a sheet with a width of w bearing a crack in its core, the amplitude of the stress intensity factor is given by:

$$\Delta K = \Delta\sigma\, F\left(\frac{a}{w}\right)\sqrt{\pi a} \qquad F\left(\frac{a}{w}\right) = \frac{1}{\sqrt{\cos\left(\pi\frac{a}{w}\right)}} \qquad \text{for} \quad \frac{a}{w} < 0.7 \;\; [5.24]$$

where $F\left(\frac{a}{w}\right)$ is the Feddersen correction factor.

To reproduce the global scattering of the experimental curves using simulation, the parameters of Paris' law [5.23] can be made random: this is the approach used from now on in this chapter. For each sample of parameters (C, m), a propagation curve is obtained and this curve can reproduce the general shape of the experimental curves. However, the modeling of the irregularities of the curves needs the Paris-Erdogan law to be modified by introducing a random process for modulating the size increments during the propagation itself. This type of approach has been studied by Ditlevsen and Olesen [DIT 86], Yang and Manning [YAN 96]. A detailed review can be obtained from [ZHE 98].

5.5.3. Probabilistic model of the data

For each crack propagation curve, the best-fit couple of parameters (C, m) may be estimated using an optimization procedure, which leads to a sample set. A statistical treatment of this sample set can then be carried out to infer the best probability density functions. Kotulski [KOT 98] shows that parameters m and $\log C$ may be reasonably represented by truncated normal distributions (Table 5.3).

Parameter	Distribution	Boundaries	Average	Variation coefficient
m	normal truncated	$[-\infty; 3.2]$	2.874	5.7%
$\log C$	normal truncated	$[-28; +\infty]$	-26.155	3.7%
Correlation coefficient: $\rho = -0.997$				

Table 5.3. *Crack propagation – probability distribution functions of* $(\log C , m)$ *of Paris law for the Virkler tests [KOT 98]* *$(da/dN$ given in mm/cycle)*

A strong correlation between both m and $\log C$ parameters can be observed (Figure 5.8), which leads us to think that a single and unique underlying parameter might exist whose variability from one specimen to another explains the dispersion of the propagation curves.

5.5.4. Propagation prediction

From the Paris model and the probabilistic description of the propagation parameters given in Table 5.3, a cluster of propagation curves can be obtained using the Monte Carlo simulation. This leads to a confidence interval (e.g. at a confidence level of 95%) regarding the crack length as a function of the number of applied loading cycles. Figure 5.9 shows the obtained median propagation curve as well as the 2.5% and 97.5% quantiles.

It appears that the scattering observed from the simulated curves has the same order of magnitude as that observed on the experimental curves, which validates the

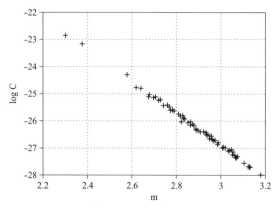

Figure 5.8. *Crack propagation – sample of the parameters of the Paris-Erdogan law for the Virkler tests (da/dN given in mm/cycle)*

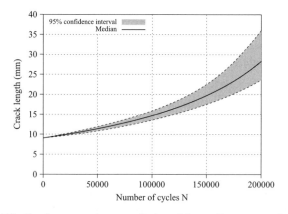

Figure 5.9. *Crack propagation – prediction of the median propagation curve and of the 95% confidence interval*

statistical treatment of the $(\log C, m)$ data previously carried out (see [BOU 08] for a detailed investigation of the influence of the correlation between $\log C$ and m on the predictions).

Nevertheless, the results presented in Figure 5.9 do not give much information in an industrial context where the best prediction of the length of the crack, as a function of the number of applied cycles, is required. The graph only leads to the conclusion that a crack with an initial size of 9 mm will reach a size ranging from 23.5 and 35.8 mm after 200,000 cycles with a probability of 95%. It is then obvious that a better estimation is required to assess, for instance, an inspection plan. The following section will focus on how to combine the previous results with the measurements performed on a structure of interest during the initial phase of propagation, in order to reduce the confidence interval of the prediction.

5.5.5. *Bayesian updating of crack propagation models*

5.5.5.1. Introduction

Bayesian statistics allows us to combine some *prior* information on the parameters of a probabilistic model with some measurement data. The reader who is unfamiliar with these concepts can get more information from Appendix A in section 5.7. In the context of probabilistic mechanics, the Bayesian framework can be applied to consistently combine, on the one hand, the predictions of a model whose uncertain input parameters are modeled by random variables, and on the other hand, the measurements of the response of the real mechanical system which was modeled. For the precise example of crack propagation, some measurements of the crack length obtained for different numbers of cycles *at the early stage of propagation* may be introduced in order to update the prior prediction which appeared inaccurate (as seen from the previous section).

5.5.5.2. Ingredients for a Bayesian approach to crack propagation

The probabilistic propagation model developed in section 5.5.4 is made of the crack propagation model (Paris-Erdogan law) and the probabilistic model of the parameters. The truncated normal distributions (Table 5.3) are considered as *prior* information (regarding the Bayesian vocabulary) on the propagation parameters in the case of 2024-T3 aluminum.

A particular specimen is now studied and the test trajectory corresponding to the slowest propagation is chosen for the sake of illustration. Figure 5.10 clearly shows that this test is singular since the propagation curve is largely outside the 95% confidence interval on the prior prediction.

To compare the observations and the predictions, a *measurement/model* error is usually defined, which considers the measurement of a physical quantity never to be perfectly accurate and the entire mathematical model of the real world to always be more or less imperfect. Thus the following link between observations and prediction is used:

$$y_{obs} = \mathcal{M}(\tilde{x}) + e \qquad\qquad [5.25]$$

where y_{obs} is the measured value and $\mathcal{M}(\tilde{x})$ is the prediction of the model for the "true value" $\tilde{x}$ of the vector of the input parameters. This true value is usually unknown, but it is assumed that it corresponds to a specific realization of vector X. The measurement/model error e is supposed to be a realization of a random variable of prescribed distribution (usually Gaussian, with mean value equal to zero and with

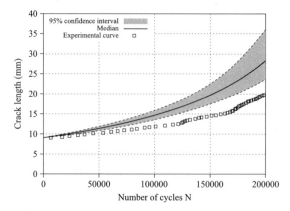

Figure 5.10. *Crack propagation – a priori prediction of the propagation (median curve and 95% confidence interval) and slowest experimental propagation curve*

standard deviation σ_e). These assumptions lead us to think that y_{obs} is a realization of a random variable Y_{obs} whose conditional distribution with respect to $\boldsymbol{X} = \tilde{\boldsymbol{x}}$ reads:

$$Y_{obs}|\boldsymbol{X} = \tilde{\boldsymbol{x}} \sim \mathcal{N}\left(\mathcal{M}(\tilde{\boldsymbol{x}}); \sigma_e^2\right) \qquad [5.26]$$

The previous equation then allows the formulation of a likelihood function for the observations and calculation of an *a posteriori* for vector $\boldsymbol{X}$ within the Bayesian framework.

5.5.5.3. *Bayesian updating methods*

The theoretical aspects of the Bayesian updating of mechanical models from observations go beyond this chapter's main topic. The interested reader can find more information in [PER 08]. To summarize, two main types of resolution methods can be distinguished regarding this issue:

– The methods which will update the *prior* probabilistic model of the input parameters of the model (here, the prior probability density functions of $\log C$ and m) from the measurements of the system response, and which then allow an *a posteriori* distribution to be estimated [PER 07b]. The propagation of this new *a posteriori* probabilistic model will allow an updated confidence interval to be calculated on the propagation curve. From an algorithmic point of view, the Markov chain Monte Carlo methods (MCMC) [ROB 96] are well adapted to the simulation of the *a posteriori* distributions.

– The methods which directly deal with the updating of the model response by defining a confidence interval, conditionally to the observations. These methods rely on the FORM approximation method that is used in structural reliability analysis [PER 07c, SUD 06].

5.5.5.4. *Application*

The evolution of the crack length during the first part of the propagation is supposed to be measured for a few values of the number of cycles (lower than 100,000). The Bayesian approach will allow the prediction to be updated, which means that a 95% confidence interval on the propagation curve will be re-estimated considering the measurement data. In this case, five measured values reported in Table 5.4 are considered.

Crack length (mm)	Number of cycles
9.4	16,345
10.0	36,673
10.4	53,883
11.0	72,556
12.0	101,080

Table 5.4. *Crack propagation measurements of the crack length on a particular sample at an early stage of propagation*

The updated median curve and the 95% confidence interval are plotted in Figure 5.11. It clearly appears that the updated prediction agrees with the observations, and that the confidence interval is significantly reduced compared to the *prior* prediction. The length of the crack at 200,000 cycles, predicted once updating has been performed, ranges from 17.4 to 19.8 mm, with a 95% probability, the measured value corresponding to the highest value.

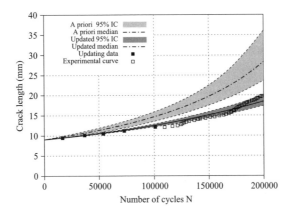

Figure 5.11. *Crack propagation – a priori and posteriori prediction of the crack propagation curve (median curve and 95% confidence interval). The black squares correspond to the measurements used for the updating; the empty squares correspond to the rest of the considered experimental curve*

It is important to note that the probabilistic formalism previously presented allows us to represent the model error: indeed, the experimental measurements of the crack length are very accurate and the measurement error can thus be considered as equal to zero. In contrast, the simplified Paris law model does not clearly allow the irregularities from the propagation curve to be reproduced. However, the model error introduced within the method (standard value $\sigma_e = 0.2$ mm for the numerical application) allows a satisfying confidence interval to be obtained. Such a result would not be obtained using the least squares fitting method, of a Paris curve on the five measurements gathered from Table 5.4.

5.5.6. *Conclusion*

In this section, the general scheme of the uncertainty modeling was applied to crack propagation using the classical Paris-Erdogan model. The observed scattering on the crack propagation rate for identical specimens was well reproduced by propagating the uncertainties identified on parameters $(\log C, m)$ through the Paris law. The 95% confidence interval tends to become very wide when the number of cycles increases. Nevertheless, considering the auscultation data (in this case, crack lengths measured at an early stage of propagation) within a Bayesian framework allows the predictions to be significantly improved as it gives an updated median curve in accordance with the observations and strongly reduces the 95% confidence interval.

The probabilistic approaches present the advantage to be used for any type of underlying physical model. Thus, the use of the extended finite element method (X-FEM) applied to crack propagation can be coupled with the methods of uncertainty modeling as presented in section 5.2, see [NES 06, NES 07].

5.6. General conclusion

The random nature of the fatigue phenomenon within materials and structures has been well known for many years. Nevertheless, the consistent and rigorous integration of all the uncertainty sources regarding the design of realistic structures is an interesting topic. In this chapter, a general methodology of the uncertainty treatment was described. This methodology can be applied to mechanics and also to any domain where the numerical simulation of physical phenomena is necessary (computational fluid, dynamics, thermal problems, neutronics, electromagnetism, chemical engineering, etc.). This general scheme can be applied to the design of the fatigue of mechanical parts; either at the crack initiation phase $((S - N)$ approach) or at the propagation step of existing cracks (Paris-Erdogan approach).

From a general point of view, a certain French cultural reluctance to use the probabilistic methods in industry can be observed, especially regarding the behavior of structures. Most of the design codes (nuclear, aerospace, civil engineering industries) are mainly deterministic (based on *conservative* design ensured by safety factors), even if some parts of the codified rules (especially the choice of characteristic values for the calculation parameters) do have some probabilistic interpretation. As fatigue analysis is a field where randomness is observed at various levels (material strength, loading, etc.) and cannot be reduced, it seems important that the uncertainty quantification methods are used as a routine in this field in the future.

5.7. Appendix A: probability theory reminder

The aim of this appendix is to recall the basics of probability theory required to understand this chapter. It is not a mathematical course on this topic. For further information on the statistical methods used in this chapter, readers should refer to the books by Saporta [SAP 06] and O'Hagan and Forster [OHA 04].

5.7.1. *Random variables*

The classical axiomatic approach of probability theory consists of building an abstract probability space characterized by the triplet $(\Omega, \mathcal{F}, \mathbb{P})$, where Ω stands for the space of outcomes, $\mathcal{F}$ stands for the σ-algebra (i.e. the subsets of Ω), and $\mathbb{P}()$ stands for probability measure. The latter allows each event A in $\mathcal{F}$ to be given a probability $\mathbb{P}(A)$ which is a real number ranging from 0 to 1.

In probabilistic engineering mechanics *random variables* (and therfore the random vectors) are used to model the uncertainty of the parameters of the mathematical model which describes the mechanical system. A random (real) variable $X(\omega)$ is defined as an application $X : \Omega \longmapsto \mathcal{D}_X \subset \mathbb{R}$. A *realization* of a random variable $x_0 \equiv X(\omega_0) \in \mathcal{D}_X$ is one of the possible values that the parameter, modeled by X, can take. The term of discrete or continuous variable is used depending on the support $\mathcal{D}_X$ (meaning the set of all the possible realizations of the variable) which can be discrete or be continuous.

A random variable $X(\omega)$ is entirely defined by its cumulative distribution function, written as $\mathcal{F}_X(x) : \mathcal{D}_X \longmapsto [0, 1]$:

$$\mathcal{F}_X(x) = \mathbb{P}(X(\omega) \leq x) \tag{5.27}$$

Therefore cumulative distribution function evaluated at point x is then the probability random variable[4] X takes values lower than or equal to x. For a continuous

---

4. From now on, the dependency on ω is omitted for the sake of simplicity. The random variables are denoted by capital letters while lowercase letters are used for realizations.

random variable, the *probability density function* is defined by:

$$f_X(x) = \frac{dF_X(x)}{dx}$$

[5.28]

Therefore, quantity $f_X(x)\,dx$ stands for the probability that X takes a value ranging from x to $x + dx$. By definition, the integral of f_X on its definition domain is equal to 1. This is also the limit of $F_X(x)$ when x tends towards the upper boundary of $\mathcal{D}_X$. For instance, the uniform distribution of an interval $[a\,,\,b]$ has a probability density function $f_U(x) = 1/(b-a)$ if $x \in [a\,,\,b]$, and 0 otherwise. A Gaussian distribution (also called normal distribution) $\mathcal{N}(\mu\ \sigma)$ is defined by the following probability density for any $x \in \mathbb{R}$:

$$f_{\mathcal{N}}(x) = \frac{1}{\sigma}\varphi\left(\frac{x-\mu}{\sigma}\right)$$

[5.29]

where $\varphi(x)$ is the standard normal probability density function which can be written as:

$$\varphi(x) = \frac{1}{\sqrt{2\,\pi}}e^{-x^2/2}$$

[5.30]

The normal cumulative distribution function, usually denoted by Φ, is defined by:

$$\Phi(x) = \int_{-\infty}^{x} \frac{1}{\sqrt{2\,\pi}}e^{-t^2/2}\,dt$$

[5.31]

It has no analytical expression, however it is tabulated in software such as Excel, Matlab, Scilab, etc.

In this chapter, the lognormal distributions are used. By definition, a random variable follows a lognormal distribution if its logarithm follows a Gaussian distribution. The following equation is then used:

$$X \sim \mathcal{LN}(\lambda, \zeta) \quad : \quad X = e^{\lambda + \zeta\,\xi} \qquad \text{with } \xi \sim \mathcal{N}(0, 1)$$

[5.32]

The probability density of a variable $\mathcal{LN}(\lambda, \zeta)$ is written as:

$$f_{\mathcal{LN}}(x) = \frac{1}{\zeta x}\varphi\left(\frac{\ln x - \lambda}{\zeta}\right)$$

[5.33]

The most likely value of a random variable corresponds to the realization x_0 which maximizes the probability density $f_X(x)$: this is what is called the *mode*. The usual distributions present a single maximum (they are called uni-modal) but some multi-modal distributions can also be built.

5.7.2. *Expected value, moments, and quantiles*

The *expected value* of a random variable is defined as:

$$E[X] \equiv \int_{\mathcal{D}_X} x \, f_X(x) \, dx \qquad\qquad [5.34]$$

This is what is usually called the *average* of X, which is also written as μ_X. As long as the integral is defined, the expected value of a function $g(X)$ can usually be defined by:

$$E[g(X)] \equiv \int_{\mathcal{D}_X} g(x) \, f_X(x) \, dx \qquad\qquad [5.35]$$

The statistical *moments* m_k (respectively centered moments μ_k) correspond to the particular case where $g(X) = X^k$ (respectively $g(X) = (X - \mu_X)^k$, $k \in \mathbb{N}$):

$$m_k = \int_{\mathcal{D}_X} x^k \, f_X(x) \, dx \qquad\qquad [5.36]$$

$$\mu_k = \int_{\mathcal{D}_X} (x - \mu_X)^k \, f_X(x) \, dx \qquad\qquad [5.37]$$

Variance $\sigma_X^2 = \mu 2 = E_{(X-\mu_X)^2}$ is the second order centered moment. The standard value of X, written as σ_X, is the square root of the variance, and the coefficient of variation (given as a percentage) is the ratio $CV_X = \sigma_{X/\mu_X}$.

The *quantiles* of a random variable are defined from the inverse cumulative distribution function. The quantile x_q with an order of q is defined by:

$$x_q : \mathbb{P}(X \le x_q) = q \qquad\qquad [5.38]$$

which can also be written as:

$$x_q = F_X^{-1}(q) \qquad\qquad [5.39]$$

if the cumulative distribution function F_X is strictly increasing.

5.7.3. *Maximum likelihood*

Step B of the general scheme of the treatment of the uncertainties consists of proposing a probabilistic model of the input parameters of the mechanical model;

that is to say, to prescribe a probability density function $f_X(x)$ for the vector X of these parameters.

In this chapter, only the case where the input variables can be considered as independent is studied : $f_X(x)$ is then the product of the marginal probability density functions f_{X_i}. If we can rely on experimental data $\mathcal{E} = \{x^{(1)}, \ldots, x^{(Q)}\}$ for the parameters, the statistical inference techniques can be used in order to determine a probability distribution that is consistent with the available data. The parametric inference consists of assuming a specific form for the probability density f_X (e.g. Gaussian, lognormal, etc.) and then in estimating the parameters of this density so that it better reproduces the sample. Let $f(X; \theta)$ be this density where $f()$ is a known function (e.g. equation [5.29]) and θ is the vector of parameters that shall be estimated (e.g. $\theta = (\mu, \sigma)$). The likelihood function of the sample set is defined as:

$$\mathsf{L}(\theta; \mathcal{E}) = \prod_{i=1}^{Q} f(x^{(i)}; \theta) \qquad [5.40]$$

Once the sample is known, this likelihood function only depends on θ. The maximum likelihood principle indicates that the best choice of the parameters (called $\hat{\theta}$) is then the one which maximizes the previous function, or equivalently minimizes the log-likelihood:

$$\hat{\theta} = \arg\min_{\theta} \left[-\log \mathsf{L}(\theta; \mathcal{E}) \right] = \arg\min_{\theta} \left[-\sum_{i=1}^{Q} \log f(x^{(i)}; \theta) \right] \qquad [5.41]$$

Usually, from the observation of sample set $\mathcal{E}$ using the tools of descriptive statistics (e.g. histograms [SAP 06, Chapter 5]), different choices are possible for $f(X; \theta)$. The best parameters are then estimated for each choice and the hypotheses are validated *a posteriori* using goodness-of-fit tests (e.g. the Kolmogorov-Smirnov test, the Anderson-Darling test, χ^2, etc. [SAP 06, Chapter 14]). The most relevant distribution can eventually be selected using likelihood criteria such as AIC (*Akaike Information Criterion*) or BIC (*Bayesian Information Criterion*) [SAP 06, Chapter 19]).

5.7.4. *Bayesian inference*

Bayesian statistics is a branch of statistics coming from Bayes' theorem which is an elementary result of the probability theory, which states that for two random events A and B:

$$\mathbb{P}(A|B) = \frac{\mathbb{P}(B|A)\,\mathbb{P}(A)}{\mathbb{P}(B)} \tag{5.42}$$

where $\mathbb{P}(A|B)$ is the *conditional probability* of A *knowing* B and is equal by definition to $\mathbb{P}(A \cap B)/\mathbb{P}(B)$. The previous result can easily be applied to the probability density functions (respectively conditional probability density functions) of random variables.

The Bayesian approach consists of integrating some *prior* information on the parameters θ to be estimated. Not only is a family of distribution $f(X; \theta)$ chosen but also a probability density function $p_\Theta(\theta)$ (called *prior distribution*) is defined for the so-called *hyper-parameters* θ. For instance the Gaussian distribution $f(X; \theta)$ with a mean value μ and a standard value σ is selected and in addition, the mean value μ is assumed to *a priori* vary between a lower and an upper bound, which comes to modeling the hyper-parameter μ by a uniform variable between these two bounds. The *a posterior distribution* f_Θ can then be deduced by combining the prior distribution and the likelihood function:

$$f''_\Theta(\theta) = c\,\mathsf{L}(\theta\,;\,\mathcal{E})\,p_\Theta(\theta) \tag{5.43}$$

where L is the likelihood function defined in equation [5.40] and c is a normalization factor which ensures that f''_Θ is a probability distribution function. This equation is similar in principle to equation [5.42] and can be read as follows: the distribution of Θ conditionally to the observations $\mathcal{E}$ is equal to the prior distribution multiplied by the likelihood function.

Focusing on the initial problem, which is to propose the best possible probabilistic model for parameter X, the mean value or the *a posteriori* mode of f''_Θ can be chosen as the best parameter.

In the case of updating the predictions of a model $Y = \mathcal{M}(X)$ by some observations $\mathcal{Y}_{obs} = \{y^{(1)}, \ldots, y^{(Q)}\}$ (see section 5.5.5), the *prior* probabilistic model is defined for X (which acts as hyper-parameters Θ in the above description) and the *posterior* distribution f''_X is eventually propagated through the physical model $\mathcal{M}$.

5.7.5. *Reliability analysis and FORM method*

A structural reliability problem is set up from the following ingredients: a mechanical model $\mathcal{M}$, a probabilistic model for its input parameters $\boldsymbol{X}$ (i.e. a probability density function $f_{\boldsymbol{X}}(x)$, and a failure criterion[5] (see [LEM 08] for a pedagogical introduction to structural reliability)). It can be mathematically formulated using a *limit state function* (also called *performance function*) $g(\boldsymbol{X})$ which shall take negative (respectively positive) values for the realizations of $\boldsymbol{X}$ such that the mechanical system fails (respectively does not fail). The set of points x such that $g(x) = 0$ defines the *limit state surface*. A common situation corresponds to the case when a quantity of interest Y (obtained as the response of a mechanical model) shall not be greater than a prescribed threshold $\bar{y}$. The associated limit state function then reads:

$$g(\boldsymbol{X}) = \bar{y} - \mathcal{M}(\boldsymbol{X}) \tag{5.44}$$

The failure probability is then given by:

$$P_f \equiv \mathbb{P}\left(g(\boldsymbol{X}) \leq 0\right) = \mathbb{P}\left(\mathcal{M}(\boldsymbol{X}) \geq \bar{y}\right) = \int_{\mathcal{D}_f} f_{\boldsymbol{X}}(\boldsymbol{x})\,d\boldsymbol{x} \tag{5.45}$$

where $\mathcal{D}_f = \{\boldsymbol{x} : g(\boldsymbol{x}) \leq 0\}$ stands for the failure domain. The Monte Carlo simulation method allows us to evaluate the above failure probability quite easily: the input random vector is *sampled*, which means that an artificial sample set of input vectors is created according to $f_{\boldsymbol{X}}$, say $\mathcal{X} = \{\boldsymbol{x}^{(1)}, \ldots, \boldsymbol{x}^{(N)}\}$. For each random vector $\boldsymbol{x}^{(i)}$ the response $\mathcal{M}(\boldsymbol{x}^{(i)})$ and then the criterion $g(\boldsymbol{x}^{(i)})$ are evaluated. The failure probability is estimated by the number of times N_f the calculation leads to a negative value of $g(\boldsymbol{x}^{(i)})$, divided by the total number of samples N:

$$\hat{P}_f = \frac{N_f}{N} \tag{5.46}$$

This method is not applicable in practice for problems where the failure probability to be evaluated is low (say from 10^{-2} to 10^{-6}): indeed, about $N = 400 \times 10^k$ realizations of X are needed to get a 5% accurate estimation of a failure probability with an order of magnitude 10^{-k}. The Monte Carlo simulation is therefore not possible, due to its cost, when each evaluation of $\mathcal{M}$ corresponds to a calculation relying on finite elements that may take several minutes or hours.

The FORM method is an approximation method which provides an estimation of P_f at low computational cost. The vector of parameters X is first transformed into a

---

5. Failure corresponds here to the "non-fulfilling of a performance assigned to the system" and not necessarily to the collapse of the considered mechanical system.

vector of standard normal random variables $X = \mathcal{T}(\xi)$, which allows equation [5.45] to be recast as:

$$P_f \equiv \mathbb{P}\left(g(\mathcal{T}(\xi)) \leq 0\right) = \int_{\mathcal{D}_f = \{\xi : g(\mathcal{T}(\xi)) \leq 0\}} (2\pi)^{-n/2}\, e^{-\|\xi\|^2/2}\, d\xi \qquad [5.47]$$

where n is the dimension of ξ. In this equation, the integrands exponentially decrease with $|\xi|^2$. Therefore, the points of the integration domain (failure domain) that contribute the most to the integral are those points that are close to the origin of this standard normal space. Thus the *design point* ξ^* is first computed, which is the point of the failure domain that is the closest to the origin of the space. The limit state surface is then linearized around ξ^*. The failure probability is then proven to be equal to:

$$P_{f,\text{FORM}} = \Phi(-\beta) \qquad [5.48]$$

where $\beta = |\xi^*|$ is the Hasofer-Lind reliability index.

For a reliability problem in which the dimension of the input vector X is lower than 10, the FORM method usually yields good results for a number of evaluations of g (which are necessary to find the design point ξ^*) lower than 100. This number is *independent* of the order of magnitude of P_f in contrast to Monte Carlo simulation. However FORM only provides an *approximation* of P_f and it is not possible to prove that this FORM result is close to the true value of P_f or not, and/or whether the result is conservative or not. Some additional techniques like *importance sampling* shall be used to validate the FORM results [LEM 05].

5.8. Bibliography

[AFC 00] AFCEN, *Règles de Conception et de Construction des Matériels des Ilots Nucléaires*, (RCCM), June 2000, Paris.

[AFN 90] AFNOR, *Produits Métalliques – Pratique des Essais de Fatigue Oligocyclique, Norme A03-403*, AFNOR, 1990.

[AFN 91] AFNOR, *Produits Métalliques – Essais de Fatigue: Traitement Statistique des Données, Norme A03-405*, AFNOR, 1991.

[ALE 04] F. ALEXANDRE, Aspects probabilistes et microstructuraux de l'amorçage des fissures de fatigue dans l'alliage Inco 718, PhD Thesis, Ecole Nationale Supérieure des Mines de Paris, 2004.

[AMZ 94] C. AMZALLAG, J.-P. GEREY, J.-L. ROBERT, J. BAHUAUD, "Standardization of the Rainflow counting method for fatigue analysis", *Int. J. Fatigue*, vol. 16, pp. 287-293, 1994.

[BAS 60] F. BASTENAIRE, Etude statistique et physique de la dispersion des résistances et des endurances à la fatigue, PhD Thesis, Faculté des Sciences de l'Université de Paris, 1960.

[BAT 10] C. BATHIAS, A. PINEAU, *Fatigue of Materials: Fundamentals*, ISTE, London and John Wiley & Sons, New York, 2010.

[BEN 06] D. BENASCIUTTI, R. TOVO, "Comparison of spectral methods for fatigue analysis of broad-band Gaussian random processes", *Prob. Eng. Mech.*, vol. 21, pp. 287-299, 2006.

[BOU 08] J.-M. BOURINET, M. LEMAIRE, "FORM sensitivities to correlation: application to fatigue crack propagation based on Virkler data", *Proc. 4th Int. ASRANet Colloquium*, Athens, 2008.

[COL 98] COLLECTIF, Re-evaluation of Fatigue Analysis Criteria, EE/S 98.317 Report, Framatome, 1998, (Final report to CEC-DG XI contract B4-3070/95/000876/MAR/C2).

[CUR 04] F. CURTIT, "INTHERPOL thermal fatigue tests", *Proc. 3rd Int. Conf. Fatigue of Reactor Components*, NEA/CSNI/R, Seville, 2004.

[CUR 05] F. CURTIT, J.-M. STÉPHAN, "Mechanical aspect concerning thermal fatigue initiation in the mixing zones piping", *Trans. 18th Int. Conf. on Struct. Mech. in Reactor Tech. (SMiRT 18)*, Beijing, China, 2005.

[DIR 85] T. DIRLIK, Application of computers in fatigue analysis, PhD thesis, University of Warwick, United Kingdom, 1985.

[DIT 86] O. DITLEVSEN, R. OLESEN, "Statistical analysis of the Virkler data on fatigue crack growth", *Eng. Fract. Mech.*, vol. 25, pp. 177-195, 1986.

[DIT 96] O. DITLEVSEN, H. MADSEN, *Structural Reliability Methods*, J. Wiley and Sons, Chichester, 1996.

[DRO 04] J. DROESBEKE, J. FINE, G. SAPORTA, *Méthodes Bayésiennes en Statistique*, Editions Technip, 2004.

[EDF 06] EDF, R&D DIVISION, Code_Aster: *Analyse des Structures et Thermo-mécanique pour des Études et des Recherches, V.7*, 2006, http://www.code-aster.org.

[GHA 91] R. GHANEM, P. SPANOS, *Stochastic Finite Elements – A Spectral Approach*, Springer Verlag, 1991.

[GUÉ 05] Z. GUÉDÉ, Approche probabiliste de la durée de vie des structures soumises à la fatigue thermique, PhD thesis, Blaise Pascal University, Clermont-Ferrand, 2005.

[GUÉ 07] Z. GUÉDÉ, B. SUDRET, M. LEMAIRE, "Life-time reliability based assessment of structures submitted to thermal fatigue", *Int. J. Fatigue*, vol. 29, no. 7, pp. 1359-1373, 2007.

[JOI 02] JOINT COMMITTEE ON STRUCTURAL SAFETY, *JCSS Probabilistic Model Code*, JCSS 2002, http://www.jcss.ethz.ch.

[KOT 98] Z. A. KOTULSKI, "On efficiency of identification of a stochastic crack propagation model based on Virkler experimental data", *Arch. Mech.*, vol. 50, no. 5, pp. 829-847, 1998.

[KRE 83] P. KREE, C. SOIZE, *Mécanique Aléatoire*, Dunod, 1983.

[LAC 97] B. LACAZE, C. MAILHES, M. MAUBOURGUET, J.-Y. TOURNERET, *Probabilités et Statistique Appliquées*, Cépaduès Editions, 1997.

[LEM 05] M. LEMAIRE, *Fiabilité des Structures – Couplage Mécano-fiabiliste Statique*, Hermès, 2005.

[LEM 08] M. LEMAIRE, "Approche probabiliste du dimensionnement – Modélisation de l'incertain et méthode de Monte-Carlo", *Techniques de l'Ingénieur, BM 5003-5004*, 2008.

[LEM 09] M. LEMAIRE, *Structural Reliability*, ISTE, London, John Wiley & Sons, New York, 2009.

[LOR 05] S. LORÉN, M. LUNDSTRÖM, "Modelling curved S–N curves", *Fatigue Fract. Engng. Mat. Struct.*, vol. 28, pp. 437-443, 2005.

[MEL 99] R.-E. MELCHERS, *Structural Reliability Analysis and Prediction*, John Wiley & Sons, 1999.

[MIN 45] M. MINER, "Cumulative damage in fatigue", *J. Appl. Mech.*, vol. 9, pp. 159-164, 1945.

[NES 06] L. NESPUREK, J.-M. BOURINET, A. GRAVOUIL, M. L., "Some approaches to improve the computational efficiency of the reliability analysis of complex crack propagation problems", *Proc. 3rd Int. ASRANet Colloquium*, Glasgow, 2006.

[NES 07] L. NESPUREK, J.-M. BOURINET, A. GRAVOUIL, M. L., "Probabilistic approach to crack propagation based on XFEM and FORM", *Proc. Fatigue Design 2007*, Senlis, France, 2007.

[OHA 04] A. O'HAGAN, J. FORSTER, *Kendall's Advanced Theory of Statistics*, Arnold, 2004.

[PAR 63] P. PARIS, F. ERDOGAN, "A critical analysis of crack propagation laws", *Trans. ASME, J. Basic Eng.*, pp. 528-534, 1963.

[PAS 97] F. PASCUAL, W. MEEKER, "Analysis of fatigue data with runouts based on a model with non constant standard deviation and a fatigue limit parameter", *J. Testing Eval.*, vol. 25, pp. 292-301, 1997.

[PAS 99] F. PASCUAL, W. MEEKER, "Estimating fatigue curves with the random fatigue-limit model", *Technometrics*, vol. 41, no. 4, pp. 277-290, 1999.

[PER 05] F. PERRIN, B. SUDRET, M. PENDOLA, M. LEMAIRE, "Comparison of two statistical treatments of fatigue test data", *Proc. Fatigue Design 2005*, Senlis, France, 2005.

[PER 06] F. PERRIN, M. PENDOLA, T. MORO, G. MORIN, J.-M. CARDONA, "Complete probabilistic framework in fatigue design: application to exhaust manifolds", in: B. TOPPING, G. MONTERO, R. MONTENEGRO (Eds.), *Proc. 8th Int. Conf. Comput. Struct. Tech.*, Las Palmas de Gran Canaria, 2006.

[PER 07a] F. PERRIN, B. SUDRET, F. CURTIT, J.-M. STÉPHAN, M. PENDOLA, "Efficient Bayesian updating of structural fatigue design curve using experimental thermal fatigue data", *Proc. Fatigue Design 2007*, Senlis, France, 2007.

[PER 07b] F. PERRIN, B. SUDRET, M. PENDOLA, "Bayesian updating of mechanical models – Application in fracture mechanics", *Proc. 18e Congrès Français de Mécanique (CFM'2007)*, Grenoble, 2007.

[PER 07c] F. PERRIN, B. SUDRET, M. PENDOLA, E. DE ROCQUIGNY, "Comparison of Markov chain Monte-Carlo simulation and a FORM-based approach for Bayesian updating of mechanical models", *Proc. 10th Int. Conf. on Applications of Stat. and Prob. in Civil Engineering (ICASP10)*, Tokyo, Japan, 2007.

[PER 08] F. PERRIN, Intégration des données expérimentales dans les modèles probabilistes de prédiction de la durée de vie des structures, PhD Thesis, Université Blaise Pascal, Clermont-Ferrand, 2008.

[ROB 92] C. ROBERT, *L'analyse statistique bayésienne*, Economica, 1992.

[ROB 96] C. ROBERT, *Méthodes de Monte Carlo par chaînes de Markov*, Economica, 1996.

[RUB 81] R.-Y. RUBINSTEIN, *Simulation and the Monte Carlo Methods*, John Wiley & Sons, 1981.

[SAL 00] A. SALTELLI, K. CHAN, E. SCOTT, Eds., *Sensitivity Analysis*, J. Wiley & Sons, 2000.

[SAL 04] A. SALTELLI, S. TARENTOLA, F. CAMPOLONGO, M. RATTO, *Sensitivity Analysis in Practice – A Guide to Assessing Scientific Models*, J. Wiley & Sons, 2004.

[SAP 06] G. SAPORTA, *Probabilités, Analyse des Données et Statistique*, Technip Publications, 2nd edition, 2006.

[SCH 06] C. SCHWOB, L. CHAMBON, F. RONDE-OUSTAU, J. BERNADOU, "Probabilistic assessment of fatigue life using a stress-based fatigue criterion", in: B. TOPPING, G. MONTERO, R. MONTENEGRO (Eds.), *Proc. 8th Int. Conf. Comput. Struct. Tech.*, Las Palmas de Gran Canaria, 2006.

[SCH 07] C. SCHWOB, Approche non locale probabiliste pour l'analyse en fatigue des zones à gradient de contrainte, doctorate thesis, Ecole des Mines d'Albi Carmaux, 2007.

[SUD 03a] B. SUDRET, Z. GUÉDÉ, P. HORNET, J.-M. STÉPHAN, M. LEMAIRE, "Probabilistic assessment of fatigue life including statistical uncertainties in the S–N curve", *Trans. 17th Int. Conf. on Struct. Mech. in Reactor Tech. (SMiRT 17)* Paper M-232, Prague, Czech Republic, 2003.

[SUD 03b] B. SUDRET, Z. GUÉDÉ, M. LEMAIRE, "Probabilistic framework for fatigue analysis", M. MAES, L. HUYSE, Eds., *Proc. 11th IFIP WG7.5 Conference on Reliability and Optimization of Structural Systems*, Banff, Canada, Balkema, Rotterdam, pp. 251-257, 2003.

[SUD 05] B. SUDRET, Z. GUÉDÉ, "Probabilistic assessment of thermal fatigue in nuclear components", *Nuc. Eng. Des.*, vol. 235, pp. 1819-1835, 2005.

[SUD 06] B. SUDRET, F. PERRIN, M. BERVEILLER, M. PENDOLA, "Bayesian updating of the long-term creep deformations in concrete containment vessels", *Proc. 3rd Int. ASRANet Colloquium*, Glasgow, United Kingdom, 2006.

[SUD 07] B. SUDRET, Uncertainty propagation and sensitivity analysis in mechanical models – Contributions to structural reliability and stochastic spectral methods, Habilitation à diriger des recherches, Université Blaise Pascal, Clermont-Ferrand, France, 2007.

[THO 99] J.-J. THOMAS, G. PERROUD, A. BIGNONNET, D. MONNET, "Fatigue design and reliability in the automotive industry", *European Structural Integrity Society*, vol. 23, pp. 1-11, 1999.

[TOV 01] R. TOVO, "On the fatigue reliability evaluation of structural components under service loading", *Int. J. Fatigue*, vol. 23, pp. 587-598, 2001.

[VIR 78] D. VIRKLER, B. M. HILLBERRY, P. GOEL, "The statistical nature of fatigue crack propagation", *Trans. ASME, J. Eng. Mat. Tech.*, vol. 101, no. 2, pp. 148-153, 1978.

[YAN 96] J. YANG, S. MANNING, "A simple second order approximation for stochastic crack growth analysis", *Eng. Fract. Mech.*, vol. 53, no. 5, pp. 677-686, 1996.

[ZHE 98] R. ZHENG, B. ELLINGWOOD, Stochastic Fatigue Crack Growth in Steel Structrues Subjected to Random Loading, Contract report ITL-98-1 for the US Army Corps of Engineers, The Johns Hopkins University, 1998.

Chapter 6

Prediction of Fatigue Crack Growth within Structures

6.1. Prediction problems

Predicting the evolution of cracks within a structure subjected to various loadings is basically a mechanical problem where the data are given by: the geometry of the structure; configuration of the initial crack; history of the forces applied; behavior law regarding the stress; and crack growth law of the material considered.

The calculation consists of finding the relationship that connects the successive configurations of the cracks to the history of the loadings.

In the case of fatigue propagation, the crack growth law gives the increase in crack length or surface per cycle, $\delta a/\delta N$, as a function of intensive variables, such as the stress intensity factors, K_i, or Rice integral, J, or the energy restitution level, Gi:

$$\frac{\delta a}{\delta N} = f(K_i \text{ or } J \text{ or } G) \qquad [6.1]$$

K_i or J or G depend on the length, a, of the cracks and on the possible variable loadings with the number of cycles, N. Therefore, the evolution of the crack at (N) is

---

Chapter written by Jean LEMAITRE.

the solution to this differential equation, which is usually nonlinear for the initial condition, $a_o(N_o)$.

The decoupling property of the static and evolution problems makes things much easier. As these intensive variables are analytically connected to the external forces, P_M:

$$K_i = K_{ri}.P_M, \quad J = J_r P_M^2, \quad G = G_r P_M^2 \qquad [6.2]$$

The reduced magnitudes, K_{ri} or J_r or G_r, can be calculated as functions of the geometry of the structure or surface of the crack for some unit forces, $K_r(a)$, $J_r(a)$ or $G_r(a)$. Some continuum mechanics calculation is involved, relying on the behavior law of the deformation of the material. Next, the crack growth law is involved (see Figure 6.1):

$$\frac{\delta a}{\delta N} = f(K_r(a)P_M(N) \quad or \quad J_r(a)P_M^2(N) \quad or \quad G_r(a)P_M^2(N)) \Rightarrow a(N) \quad [6.3]$$

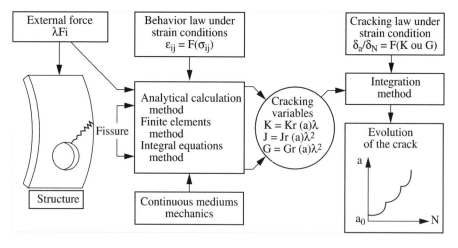

Figure 6.1. *Prediction problem flowchart of the evolution of a crack*

Knowing the failure mechanics and means of calculation does not always lead to the resolution of the entire prediction problem (3D inelastic structure subjected to a loading with variable intensity and direction) but many industrial problems correspond to some simplifying hypotheses that do give a complete resolution.

6.1.1. *Structures*

The treatment of the surface cracks in 3D is a complex problem. This is because the shape of the crack front is represented by an unknown value and its increase is represented by a vector field. If the cracked area of the structure can be considered as 2D, the crack is then linear and its growth is represented by a single vector that comes to a scalar value in the frequent cases where the crack spreads in the direction of its extension (see Figure 6.2). It is worth mentioning that under some specific loading conditions, thin structures do not depend on the plane stress hypothesis, whereas thick structures can be analyzed using the plane strain hypothesis.

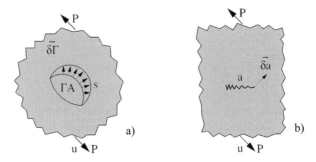

Figure 6.2. *a) Surface crack within three dimensions;*
b) linear crack within two dimensions

6.1.2. *Loading*

If the type of loading changes or the direction of the external forces change, the crack no longer grows within its plane – a bifurcation occurs. For structures that do not present any sudden geometry variations, such as holes or over-thicknesses close to the crack, the simple loading condition regarding plasticity is a non-bifurcation condition. It approximately occurs if all of the external forces proportionally increase with a single parameter.

Each external loading can then be written as the product from a scalar value that only depends on time t by a vector, which depends on the application point of force M alone:

$$\overrightarrow{P}(M,t) = \lambda(t)\,\overrightarrow{P}(M) \tag{6.4}$$

With the variation of forces as a function of time, we can distinguish:

– the nominal forces whose cracking laws are well established;

– the forces with a complex history that lead to the consideration of memory effects;

– the random forces that are only known by some statistical magnitudes. The nonlinear property of the cracking law makes the prediction hard to calculate.

6.1.3. *Material*

With the calculations of some structures, where it is necessary to know the stress intensity factors or the energy restitution level, the materials are represented by their behavior law against deformation:

– elastic and linear;

– elasto-plastic;

– elasto-visco-plastic.

In most practical cases, the whole structure operates within the elastic domain. The plastic zone at the crack tip can be neglected and calculation can be performed under elastic conditions.

Before focusing on the corresponding calculation techniques, the form of the cracking laws will be studied so the precise nature of the variables can be calculated.

6.2. Crack growth laws

6.2.1. *Mechanical definition of the cracking variables*

Three variables are usually used to characterize the cracking behavior of materials. They are equivalent in the simple case of the 2D elastic linear and fragile solids under simple loading, but can lead to some relatively high degrees of generalization for the most complex cases.

6.2.1.1. *Stress intensity factors*

These are usually defined, as presented above, as quantities characterizing the singularity of the stress field close to the crack tip. A more interesting definition involves introducing them due to the displacement field of the crack opening [BUI 68].

If the crack is repeated with coordinates (O, r, θ), within the systems with axis (O, x_1, x_2), three simple opening models, u, of the crack are defined (see Figure 6.3):

– mode I:

$U_2 \neq 0,$

$U_1 = 0,$

$U_3 = 0;$

– mode II:

$U_1 \neq 0,$

$U_2 = 0,$

$U_3 = 0;$

– mode III:

$U_3 \neq 0,$

$U_1 = 0,$

$U_2 = 0.$

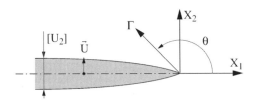

Figure 6.3. *Crack opening displacement*

The displacement field close to the crack tip is calculated using the Westergaard method [LIE 68a, LIE 68b]. For each of the three models, this is given by:

$$u_1 = \frac{K_I}{2\mu}\left[\frac{r}{2\pi}\right]^{1/2} \cos\frac{\theta}{2}[K - \cos\theta] + \frac{K_{II}}{2\mu}\left[\frac{r}{2\pi}\right]^{1/2} \sin\frac{\theta}{2}[k + \cos\theta + 2]$$

$$u_2 = \frac{K_I}{2\mu}\left[\frac{r}{2\pi}\right]^{1/2} \sin\frac{\theta}{2}[K - \cos\theta] + \frac{K_{II}}{2\mu}\left[\frac{r}{2\pi}\right]^{1/2} \cos\frac{\theta}{2}[k + \cos\theta - 2] \quad [6.5]$$

$$u_3 = K_{III}\frac{2}{\mu}\left[\frac{r}{2\pi}\right]^{1/2} \sin\frac{\theta}{2}$$

where μ is the Lam, é elasticity constant, $(3 - v)/(1+v)$ under plane stress, and $k = 3 - 4v$ under plane strain, v being Poisson's elasticity coefficient.

K_I, K_{II} and K_{III} are the stress intensity factors corresponding to modes I, II and III. They can be written as a function of the discontinuity of the crack opening displacements:

$$[u_i] = u_i(r,\pi) - u_\varepsilon(r,-\pi) \qquad [6.6]$$

mode I:

$$[u_2] = \frac{K_I}{\mu}\left[\frac{r}{2\pi}\right]^{1/2}[k(v)+1)]$$

mode II:

$$[u_1] = \frac{K_{II}}{\mu}\left[\frac{r}{2\pi}\right]^{1/2}[k(v)+1)]$$

mode III:

$$[u_3] = \frac{4K_{III}}{\mu}\left[\frac{r}{2\pi}\right]^{1/2}$$

6.2.1.2. Rice and Bui's integrals

A less local approach consists of defining the outline integrals that give the variation in potential energy of the solid for growth of the crack. Still considering a plane issue under linear elasticity conditions, Rice's integral [LIE 68a, LIE 68b] and its dual, Bui's integral [BUI 68], respectively, are:

$$J = \int_r \left[wn_1 - \sigma_{ij}n_j\frac{\partial u_i}{\partial x_1}\right]ds$$

$$[6.7]$$

$$I = \int_r \left[-w * n_1 - u_i n_j\frac{\partial \sigma_{ij}}{\partial x_1}\right]ds$$

Γ is an open linear integration outline, surrounding the crack (see Figure 6.4), integrals J and I being independent of the dimensions and norm of this outline. We use the summation convention on Einstein's dummy indices.

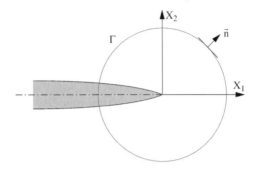

Figure 6.4. *Integration outline for Rice and Bui's integrals*

Where $\vec{n}$ is normal to Γ, w is strain energy, w^* complementary energy, σ_{ij} and ε_{ij} are components of stress and strain:

$$\sigma_{ij} = \frac{\partial w}{\partial \varepsilon_{ij}}, \quad \varepsilon_{ij} = \frac{\partial w^*}{\partial \sigma_{ij}} \qquad [6.8]$$

By applying the expressions of J and I to the Westergaard analysis, it can be shown that, regarding the plane problems:

$$J = I = \frac{1}{E}\left[K_I^2 + K_{II}^2\right] \text{ under plane stress conditions} \qquad [6.9a]$$

$$J = I = \frac{1-v^2}{E}\left[K_I^2 + K_{II}^2\right] \text{ under plane strain conditions} \qquad [6.9b]$$

6.2.1.3. *Energy restitution rate*

The energy restitution rate is a more global approach that gives the elastic energy transfer released by the solid cracked during the growth of the crack (G) and capable of creating this growth by decohesion of the material. The main interest of this concept, initially introduced by Griffith [GRI 20], is that it leads to a generalization of the linear fracture mechanics to the 3D and nonlinear problems in the case of the elasto-plastic or elasto-visco-plastic behaviors of materials [LEM 76]. The thermodynamics of irreversible processes provides the frame of this generalization, considering that area A of the crack is an internal variable hidden within the cracked solid that can be connected to the dual variable G, which then derivates from the thermodynamic potential, ψ [LEM 78]:

$$G = -\frac{\partial \psi}{\partial A}$$

[6.10]

In the simple case of an external single force, P, which corresponds to an elastic displacement, u_e (see Figure 6.2) under isothermal conditions, ψ is the energy of the elastic deformation. ψ can be written as a function of the stiffness R of the structure, which is also given as a function of the surface of the crack:

$$R = \frac{P}{u_e}, \quad \psi = \frac{1}{2} R(A) u_e^2$$

[6.11]

$$G = -\frac{1}{2} u_e^2 \frac{dR}{dA}$$

This equation directly leads to the "compliance method", allowing us to experimentally determine the rate of energy restitution of a cracked structure for external loading corresponding to the elastic displacement, u_e.

In addition, parameter G can also be given as a function of variation of the work of external forces to obtain equations that can be directly used with numerical calculation. Both the static and kinetic approaches provide a frame of the real value of G, which is what happens with the calculation of the structures. The kinetics of the growth mechanism of the crack might then occur in two different ways (see Figure 6.5):

– an imposed constant force, P, and a variable displacement, u_e, lead to the static approach:

$$R = \frac{P}{u_e}, \quad P = Const \rightarrow \frac{\delta R}{\delta A} = -\frac{P}{u_e^2} \frac{\delta u_e}{\delta A}$$

[6.12]

$$G = -\frac{1}{2} u_e^2 \frac{\delta R}{\delta A} = \frac{1}{2} P \frac{\delta u_e}{\delta A}$$

– an imposed constant displacement, u_e, and a variable force, P, lead to the kinetic approach:

$$R = \frac{P}{u_e}, \quad u_e = Const \rightarrow \frac{\delta R}{\delta A} = \frac{1}{u_e}\frac{\delta P}{\delta A} \qquad [6.13]$$

$$G = -\frac{1}{2}u_e^2\frac{\delta R}{\delta A} = -\frac{1}{2}u_e\frac{\delta P}{\delta A}$$

In both cases, the energy restitution rate is equal to half the work variation of the external forces, w, for an increase of the crack's surface.

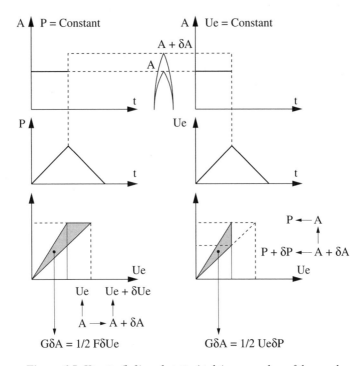

Figure 6.5. *Kinetic (left) and static (right) approaches of the crack growth mechanism*

The static approach minimizes the variation of the potential energy. It therefore maximizes the variation of the work of the external forces. The reverse phenomenon occurs in the case of the kinetic approach:

$$\frac{1}{2}\frac{\delta W}{\delta A}(P = Const) > G > \frac{1}{2}\frac{\delta W}{\delta A}(u = Const) \qquad [6.14]$$

6.2.2. *Fatigue crack growth laws*

Every fatigue law gives the growth in length or surface of the crack per cycle, $\delta a/\delta N$, as a function of cracking variables, K_i, or G. The different equations proposed are relatively complex depending on whether or not they give the effect of the average loading, of the history effect, etc. They are actually phenomenological equations whose coefficients have to be identified according to the experiments run on every material. Therefore, they are usually only valid within the variation domain of the parameters covered by the experimental results. During the experiments, there is always a gap between these results and the mathematical model. The value of the specific coefficients depends on the chosen operating domain. Depending on the different applications performed, the calculation of these coefficients can be much more accurate for each particular case within the estimated $\delta a/\delta N$ domain rate. This is much more interesting than reading the tables, which do not usually give any information on how the values could be obtained.

6.2.2.1. *Crack growth laws under nominal loadings*

This section focuses on the linear elastic 2D solid under simple loading in mode I. As the external loadings are recurrent, the stress intensity factor, K_i, which from now on is written as K (or J or G) is an increasing function of the crack length. The effect of the plastic zone, which occurs at the crack tip, can be neglected.

The basic law is Paris' law [6.15]:

$$\frac{\delta a}{\delta N} = C_1 [\Delta K]^{\eta_1} \qquad\qquad [6.15]$$

with:

$$\Delta K = K_M - K_m = K(P_{\max}) - K(P_{\min})$$

where C_1 and η_1 are coefficients that are specific to the material to be determined thanks to basic experiments on CT (compact tension) or Irwin specimens [LIE 69a, LIE 69b, LIE 72] (η_1 is of about second to fifth order for common metallic materials). The validity domain of this law, which does not consider the effect of the static load, is limited to cases where $K_m \simeq 0$. The average loading effect is introduced by the ratio $R = K_m/K_M$; Forman's law [LIE 71] presents this variable and, in addition, allows the fragile failure to be determined using:

$$(K_M \to K_c) \Rightarrow \frac{\delta a}{\delta N} \to \infty \qquad\qquad [6.16a]$$

$$\frac{\delta a}{\delta N} = \frac{C_2 \Delta K^{\eta_2}}{[1-R][K_c - K_M]}$$

bearing three parameters specific to the material: C_2, η_2 and K_c.

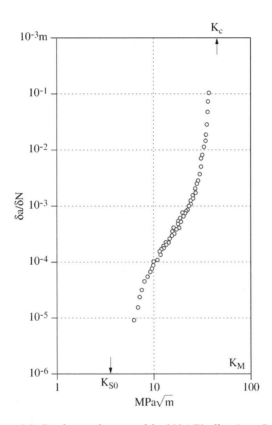

Figure 6.6. *Crack growth curve of the 2024-T3 alloy (tests $R \approx 0$)*

Another way to consider the existence of the non-cracking threshold, K_{so}, is the following equation [PEL 77a, PEL 77b, PEL 77c]:

$$\frac{\delta a}{\delta N} = C_3 \left[\frac{K_M \dfrac{1-R}{1-mR} - K_{so}}{K_c - K_M} \right]^{\eta_3} \qquad [6.16b]$$

bearing five parameters C_3, η_3, K_{so}, K_c and m. The first can be identified from a typical graph, such as that in Figure 6.6. Coefficient m (close to 0.5) can be obtained from some test results, $a(N)$, with different average loadings.

Many other mathematical models have been proposed but they only slightly improve on these three basic models.

In the case of mixed modes, the most simple and accurate method consists of using these laws with the energy restitution rate, G, for instance:

$$\frac{\delta A}{\delta N} = C_4 G^{\eta_4}$$

[6.17]

A this time being the surface of the crack (the variable connected to G), which in 3D is equal to $e. a$ (e being the thickness of the medium).

6.2.2.2. *Crack growth laws under random loadings*

When loading is no longer constant, the plastic zone, or to be more exact the geometrical zone within which plastic dissipations occur, is where a memory effect due to the irreversibility of the plastic deformations can be observed [SCH 73, WEI 76]. The dimensions of the plastic zone depend on the current loading but also on strong loadings from the recent past. These lead to the idea of introducing a variable specific to the dimensions of the plastic zone to explain this history effect.

Due to some 2D problems under a simple loading (but with a complex history) and a simple mode, the structure is now considered linear elastic and locally plastic. The low dimension of the plastic zone involving the dimension of the crack allows a simplified calculation of the plastic zone to be carried out. This is quite useful as no calculation under elasto-plastic conditions is required.

The length, ρ, of the plastic zone at the tip of a crack loaded for the first time is the distance from the crack tip, whose normal strain σ_{22} calculated under elasticity conditions is equal to the elasticity limit of the material σ_y. It usually leads to the following length:

$$\rho_k = \frac{K_I^2}{2\pi\sigma_y^2}$$

[6.18]

a_s being the crack length corresponding to the last over-loading P_{MS} (see Figure 6.7).

The Paris-Wheeler crack-growth law can be written as:

$$\frac{\delta a}{\delta N} = c_5 \left[k_M \frac{1-R}{1-mR} \right]^{\eta_s} \left[\frac{\rho_K}{\rho_N} \right]^{\beta_1}$$

[6.19]

bearing four parameters: C_5, η_5, m and β_1.

Coefficient β_1 can be identified from the basic experimental results with over-loadings to show the growth delay, such as those presented in Figure 6.8.

From a phenomenological point of view, the memory effect can be understood thanks to the values of the residual stresses once the over-loading has been applied. This is the Willenborg model [WIL 71], whose validity domain is equivalent to the Wheeler model.

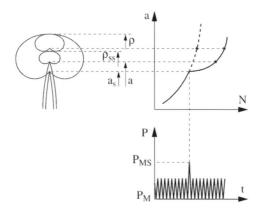

Figure 6.7. *Wheeler's model*

The Wheeler model is only valid in the case of isolated over-loadings. If the loading history is more complex, a more complicated model has to be used to explain what really occurs. From a thermodynamic point of view, the state of the elastic and locally plastic cracked structure can be represented by two internal values: the crack length a, which corresponds to the energy restitution ratio G; and the dimension of the plastic zone ρ_N, which is related to the dual variable of the threshold energy restitution ratio G_s. Therefore, ρ_N is derived from the plastic dissipation quasi-potential φ_p [LEM 78]:

$$\rho_N = \frac{\partial \varphi_p}{\partial G_s}$$

[6.20]

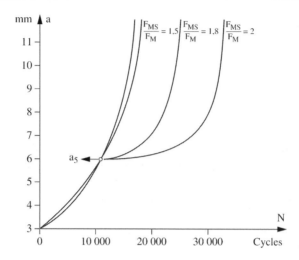

Figure 6.8. *Cracking tests with an overloaded 2024-T3 alloy.*
Model with two variables: K_m and K_s

When this is transposed to the variable representing the stress intensity factor, two differential equations are then formulated [PEL 77a, PEL 77b, PEL 77c]:

$$\frac{\delta a}{\delta N} = C_6 \left[K - K_s \right]^{\eta_6}$$

$$\frac{\delta K_s}{\delta N} = \alpha(R) \left[\frac{K_M}{R_2(R)} - K_s \right] \left[\frac{\rho_K}{\rho_N} \right]^{\beta_2}$$

[6.21]

K_s is the variable representing the intensity factor of the instantaneous non-cracking threshold stress, which depends on the loadings previously applied. It is equal to K_{so}, the initial threshold, if and only if the crack has not been subjected to any loading in the past. This is related to the concept of crack opening and closure [ELB 71].

C_6, η_6, $\alpha(R)$, R_2 (R) and β_2 are coefficients or functions that have to be determined for every material, according to the basic tests with constant amplitude and subjected to an over-loading.

6.2.2.2.1. RMS laws

In the case of random loadings, a rough but useful approximation justified by many tests [ROL 77] consists of correlating the average cracking rate, $\overline{\delta a / \delta N}$, to the quadratic average, ΔK_{RMS}, of the amplitude of the stress intensity factor:

$$\frac{\overline{\delta a}}{\delta N} = C_7 \Delta K_{RMS}^{\eta_7}$$ [6.22]

with:

$$\Delta K_{RMS} = \left[\frac{1}{k} \sum_{i=1}^{k} \Delta K_i^2 \right]^{1/2}$$

Obviously, number k has to be high enough for this average to be represented thanks to a stationary process, but low enough to consider the variation, K, along with the crack length, a.

This model gives some good results for Gaussian loadings as a function of time. The amplitude of these loadings depends then on a Rayleigh probability density [CRA 63]. The results are much less satisfactory in the case of a loading history presenting numerous over-loadings with high amplitudes. The model bearing two variables, K_M and K_s, is therefore preferred.

6.2.3. Bifurcation criteria

When the direction of loading varies (non-simple loading), the direction of the crack changes. It bifurcates, which means that its growth has to be represented by a vector and no longer by a scalar variable (see Figure 6.9). A criterion setting the direction has to be added to the crack growth laws: a bifurcation criterion.

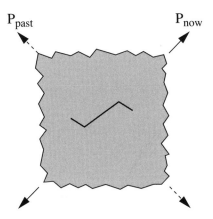

Figure 6.9. *Bifurcation of a crack*

Among the different proposed criteria, three can easily be used to make the calculation simpler:

– *Maximum normal stress criterion* [ERD 63]: the crack growth vector is normal to the direction of the highest main stress. This criterion, which can be easily applied, is only valid for low bifurcation angles.

– *Maximum and local energy density*: according to the criterion proposed by Sih [SIH 72], the journey of the crack tip can be identified thanks to the location of points of the structure that bear the maximum elastic energy density, w:

$$W = \frac{1}{2}\sigma_{ij}\varepsilon_{ij} \tag{6.23}$$

These two criteria are challenging from a numerical point of view as the direction of the highest main stress or maximum density point of the elastic energy depend on the distance from the targeted point to the crack tip.

– *Maximum energy restitution ratio criterion* [LEM 76]: among every virtual crack growth, $\overrightarrow{\delta a}*$, kinetically acceptable and with the same length, δa, the real growth, $\overrightarrow{\delta a}$, is the one that maximizes the energy restitution level, G:

$$\delta a_{real} \rightarrow \sup_{\overrightarrow{\delta a}*}\left(G(\overrightarrow{\delta a}*) \right) \tag{6.24}$$

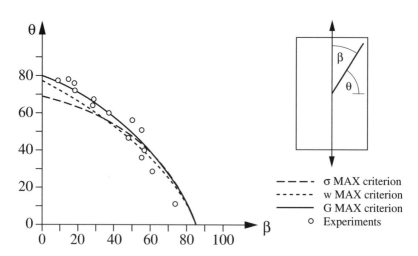

Figure 6.10. *Bifurcation criteria*

Figure 6.10 compares the prediction experiment of a bifurcation with an angle θ in the case of a plane panel where the crack has an angle β with the applied force. In this example (at least), the criterion of maximum restitution energy rate gives the best results.

6.2.4. *Three-dimensional elements*

Both the global and local approaches lead to some different prediction methods, without resolving any fundamental issue.

6.2.4.1. *Global approach*

If the crack can be represented by a small number of parameters, for instance two for an elliptical crack that remains elliptical, the energy restitution rate can be generalized [BOI 77, LEM 76]. The surface growth of a crack is given by:

$$\frac{\delta A}{\delta N} = C_4 G^{n_4} \tag{6.25}$$

The shape of the crack front comes from an optimization of the parameters that define the crack. Among all the crack growths that are kinetically acceptable, $\delta \Gamma^*$, (see Figure 6.2) with the same surface, δA, the real growth, $\delta \Gamma$, is the one that maximizes:

$$G = \frac{1}{2} \frac{\delta W}{\delta A} \tag{6.26}$$

This concept is taken in the case of a crack that is represented by an infinite number of parameters. It leads to focusing the whole crack growth at the point where the stress intensity factor is the highest [ESC 78].

6.2.4.2. *Local approach*

Another method consists of calculating the stress intensity factor within the planes that are normal to the crack tip as a function of the curved abscissa axis s (see Figure 6.2) and of applying a linear law, for instance, Paris' law [BRO 74]:

$$\frac{\delta a(s)}{\delta N} = C_1 \left[\Delta K(s) \right]^{n_1} \tag{6.27}$$

This method, which discards the 3D character of the problem, is hard to justify.

Strifors [STR 74] proposed a local energy concept and a generalization, based on the normality law of the thermodynamics of the irreversible processes [NGU 78].

If $\gamma(s)$ represents every parameter defining a growth of the crack tip, $\delta\Gamma$ (see Figure 6.2) and $\varsigma(s)$, the associated variable such as the dissipated ω is:

$$\omega = \int_{\Gamma} \varsigma(s).\gamma(s)ds = G.A \qquad [6.28]$$

The existence of a quasi-potential dissipation identified from ω leads to the following formula:

$$\varsigma(s) = \frac{\delta\omega}{\delta\gamma(s)} \qquad [6.29]$$

The cracking law can be written as an equation close to the Prandtl-Reuss plasticity law [LAB 78a, LAB 78b]:

$$\gamma(s) = \alpha\left(\varsigma^T.\varsigma\right).\varsigma(s) \qquad [6.30]$$

In this equation, T stands for the transposition. The numerical operating system of this law consists of a discretization of the crack tip; its growth is represented by a set of vectors, γ_i. Here ς is the work gradient of the external forces regarding these variables, γ_i. α is a scalar variable that can be obtained thanks to the identification of this equation with Paris' law or any other cracking model.

6.3. Calculation of cracking variables

The crack growth laws can be given as a function of the stress intensity factor variables, K_I, of Rice's integral J or of the energy restitution ratio G. As these variables are decoupled from the evolution time of the crack, their reduced values can then be calculated (for the external unit forces) without having to simultaneously resolve the cracking law. We only need to know the spatial journey of the crack. It is worth mentioning the definition of the reduced variables for a problem dealing with a simple loading where the external forces can be written as:

$$\vec{P}(M,t) = \lambda(t)\,\vec{P}(M) \qquad [6.31]$$

$$K_r = K\left(\lambda = 1\right)$$

$$J_r = J\left(\lambda = 1\right)$$

$$G_r = G\left(\lambda = 1\right)$$

6.3.1. *Two-dimensional problems under linear elasticity and simple loading conditions*

6.3.1.1. *Analytical solutions*

A certain number of elasticity issues corresponding to typical cases of cracked solids were resolved, either entirely or from a side approach, using methods based on specific analytical functions or on the standard representation. The stress intensity factors defined by the singularity of the stress field are given below for different specific cases drawn from the literature [BRO 74, BUI 78, SIH 73, SIH 77a, SIH 77b].

6.3.1.2. *Finite element method*

This method consists of turning the structure into elements within which the displacement field is approximated by simple functions whose unknown parameters are linearly connected to the displacement of the element nodes. Writing the general equilibrium of the structure as a variational equation, the linear system to be resolved can be obtained [ZIE 73].

The specific problem that appears regarding cracks is the choice of the mesh of elements close to the crack tip. The elements have to be really small due to the strong stress gradient. A way to obtain bigger elements, within the rest of the structure, has to be found in order to limit the order of the system to be resolved.

Three procedures were used:

– *Mesh into "cobwebs"* (see Figure 6.11 [CHA 74]): this breakdown gives the best precision but some work has to be done to go from one crack length to another.

– *Constant space mesh* (see Figure 6.11, [BRO 74]): compared to the previous procedure, there is no way some elements that small can be obtained. It is therefore less accurate to go from one crack length to another. This is because the bonds at the crack nodes have to be broken.

– *Use of a specific "crack tip" element*, which bears the singularity.

Thanks to the finite element method, the stress intensity factors are not usually accurately calculated using the singularity of the stress field. Better precision is obtained by using the definition of K_I from the opening field of the crack by the equations given in section 6.2.1.1.

For a crack length, a, using discreet values and for a unit stress calculation gives the displacement of the crack opening as a function of the distance from the crack tip, r:

$$[u_i](r) = u_i(r, \pi) = u_i(r_i, -\pi) \qquad [6.32]$$

and, for instance, when mode I is used the following equation is obtained:

$$K_{rI} = \lim_{r \to 0} \left(\frac{[u_2](r)}{\sqrt{\dfrac{r}{2\pi}}} \frac{\mu}{k(v)+1} \right)$$

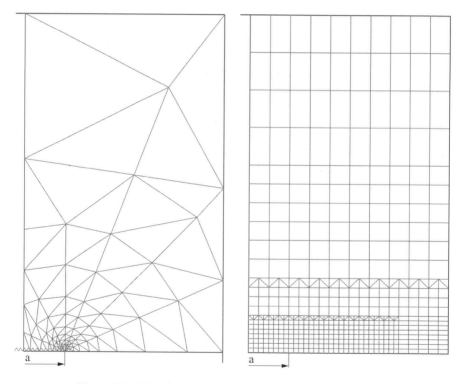

Figure 6.11. *Two discretizations of the quarters of cracked sheets*

A method that gives better and valid results under a mixed mode [CEM 77] consists of calculating the energy restitution ratio by the work variation of the external forces:

$$G_{R1} \hspace{10cm} [6.33]$$

whose evolution has to be determined as a function of the length of the crack, a:

$$G_r(a)$$

For a 2D medium with a thickness, e:

$$\delta A = e\delta a$$

If P_j are the external forces applied to the external nodes of the structure corresponding to the unit loading $\lambda = 1$, and if μ_j are the corresponding displacements, both the static and kinetic methods can be applied, as will be discussed below. Usually 10–20 crack lengths are enough to get an acceptable definition of function $G_r(a)$. Therefore, 10–20 different calculations have to be carried out.

6.3.1.2.1. Static method

For constant external forces with $\lambda = 1$, the external work, W, is calculated thanks to:

$$W(a) = P_j u_j(a)$$

(with the summation convention on subscript j).

$U_j(a)$ being the displacements calculated for different crack lengths, a, G_r can be obtained once function $W(a)$ has been smoothed or finite differences have been used:

$$G_r(a) = \frac{1}{2}\frac{dW}{dA} \hspace{8cm} [6.34]$$

6.3.1.2.2. Kinetic method

If the external forces are still displayed, and not the displacements, for each crack length, two calculations have to be performed:

– a first calculation, such as the previous ones or one from forces $P_j(a)$ and displacements u_j can be obtained for the crack length, a (static method);

– a second calculation corresponding to a crack length $a + \delta a$ with displacements U_j as reference data. This calculation gives forces $P_j(a + \delta a)$.

G_r can be obtained thanks to:

$$G_r = \frac{1}{2} \frac{u_j P_j (a + \delta a) \ - u_j P_j (a)}{\delta A}$$ [6.35]

If the stress intensity factor has to be identified, here is what has to be calculated under mode I conditions:

$$K_r = \sqrt{EG_r} \text{ , under plane stress conditions}$$ [6.36a]

$$K_r = \sqrt{\frac{E}{1-v^2} G_r} \text{ , under plane strain conditions}$$ [6.36b]

Some variation examples of the stress intensity factor, K_i, within stiff cracked thin sheets, calculated using the static method, are given in Figure 6.12 [CHA 74].

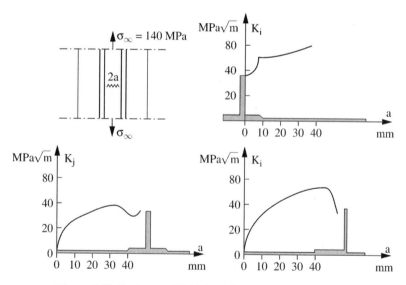

Figure 6.12. *Variation of $K_i(a)$ for different types of stiff sheets made from a 2618A-T6 alloy*

The finite element method can also be used to calculate integral J (or integral I) if and only if the integration outline is far enough from the crack tip [BUI 78]. The equation of the definition of J then has to be applied, and the variation of the quantities to be integrated on the outline is given by the interpolation formulae of the finite elements. From this calculation scheme, there are no numerical methods that allow the number of memories within the computer to be minimized and also allow the calculation times to be reduced:

– the perturbation method [HEL 75], which allows the variation of work of the external forces δW to be calculated by only using a single perturbation of the stiffness matrix. This only involves the few elements close to the crack tip being displaced in order to get a growth equal to δa;

– the reverse flexibility method [LAB 78a, LAB 78b], which also allows the calculation time to be reduced by writing the stiffness matrix in such a way that a single sub-matrix varies when the crack length varies.

6.3.2. Three-dimensional problems

6.3.2.1. Analytical solutions

Once again, a certain number of typical problems can be resolved using the elasticity theory. Most of them are described in reference [KAS 75]. Some examples are listed below:

– semi-infinite medium with a semi-elliptical crack:

$$K_{I\ max} \approx 1.12 \frac{\sigma\sqrt{\pi a}}{\frac{3\pi}{8}+\frac{\pi}{8}\frac{a^2}{c^2}} \qquad [6.37]$$

$$K_{I\ min} \approx 1.12 \frac{\sigma\sqrt{\pi a^2 / C}}{\frac{3\pi}{8}+\frac{\pi}{8}\frac{a^2}{c^2}}$$

– infinite medium with an elliptical crack:

$$K_i(\theta) = \frac{\sigma}{E(h)} \rightleftharpoons \sqrt{\frac{b}{a}} \rightleftharpoons \left[a^2\sin^2\theta + b^2\cos^2\theta\right]^{1/4}$$

$$E(h) = \int_0^{\pi/2} \left[1-h^2\sin^2 u\right]^{1/2} du \qquad [6.38]$$

$$h = \left[1-\sqrt{b/a}\right]^{1/2}$$

6.3.2.2. *Integral equations method*

For cracked structures with a complex geometry, numerical analysis has to be used. The finite element method can be used now, but handling 3D elements is not that easy to do. The integral equations method is, in this case, the best option [LAC 76]. Under elastic conditions, it consists of a transformation of the Navier equation into a surface integral that connects displacements and tensions at every point of the external outline of the structure, with some of them being data. The resolution of this equation therefore gives unknown tensions and displacements; some simple operating equations allow the deformation stress to be identified in any point of the internal domain. The problem of dealing with three dimensions in space is reduced to a 2D problem with a surface integral. This greatly simplifies the discretization, allows the data to be collected easily and makes the analysis of the results simpler.

Within the calculation programs of the computer, using the subdivision technique of the structure split into "sub-regions" is quite handy because every sub-region has its own integral equation described. Within the reference program [BOI 77] the discretization of the structure is carried out using some quadrilateral elements with eight degrees of freedom.

Using this method for the calculation of variables K_i, J or G is similar (if transposed to the 3D reference) to the one presented for the finite element method in section 6.3.1.2.

6.3.2.3. *Weight function method*

The Bueckner weight function [BUE 70] gives the stress intensity factor at a specific point of the crack tip for a force that is focused on a random spot of the structure. This function can be tabulated, once and for all, for cracked structure geometry and used for various types of superposition loadings. In order to obtain the weight function, it appears that the cheapest method is actually to calculate, through digital means, the related Green function [BUI 78].

6.3.3. *Issues under plastic and visco-plastic conditions*

If the dimensions of the plastic zone at the crack tip are too high compared to the length of the crack, or if the entire structure is loaded beyond the elasticity domain of the material, the dissipated energy under a plastic deformation can no longer be neglected. Fortunately, the energy restitution rate G can be written as a function of the elastic stiffness of the structure if the (simple) loading does not lead to any unilateral conditions. This stiffness can be obtained using a calculation under elastic conditions, ignoring the plasticity or visco-plasticity. Many problems fit into this category as long as there is no local partial reloading during the unloading of the external forces applied to the structures. This extension of the variable of the energy restitution rate [LEM 76] seems to be the easiest way to describe plasticity and visco-plasticity issues.

NOTE.– Some elements can be found in the numerical analysis techniques in [LUX 77], in the dynamic issues in [BUI 78, SIH 77a, SIH 77b] and in the thermo-elastic problems in [BUI 78].

6.4. Resolution method of the cracking equations

As the analysis of the structure leads to the calculation of $K_{rI}(a)$, $J_r(a)$ or $G_r(a)$ for external unit forces, magnitudes $K_i = K_r\lambda(N)$, $J = J_r\lambda^2(N)$ or $G = G_r\lambda^2(N)$ now have now to be incorporated in the cracking law in order to understand the evolution of the crack length $a(N)$ generated by the loading program given by $\lambda(N)$.

6.4.1. Identification of the crack growth laws

The first step in identifying crack growth laws consists of finding the typical coefficients of the material that correspond to the case being studied. The crack growth laws are phenomenological, that is to say some mathematical models approximately represent the physics without defining it very precisely. In order to improve the prediction, identification is best done in the lowest variation domain of the variables, corresponding to the estimated variation domain of the problem's variables. Figure 6.6 shows that exponent η_I of the Paris' law can be multiplied by up to three times if the identification step is performed within the rate range:

$$10^{-4} < \frac{\delta a}{\delta N} < 10^{-3}\,\mathbf{mm\,/\,c} \ \ \text{or} \ \ 10^{-3} < \frac{\delta a}{\delta N} < 10^{-2}\,\mathbf{mm\,/\,c} \qquad [6.39]$$

The configuration parameters of the problem that exactly or approximately define the identification domain are as follows:

– 2D problem under plane stress or under plane strain: in order to get rid of the thickness effect, the basic tests performed on the specimens whose thickness is close to that of the studied case should always be used.

– 3D problem: using some tests performed on the "thick" parts of a structure.

– Temperature and frequency: once again, in order to get rid of a possible effect of temperature or frequency, some basic tests run at the operating temperature of the structure and at the estimated frequency of the loading, are best used.

– *Domain of the crack growth rate $\delta a/\delta N$*: most of the time, the targeted or estimated lifetime of the structure is in the rate range to be defined (where the prediction has to as accurate as possible):

- low rates 10^{-7} mm/c $< \delta a/\delta N < 10^{-5}$ mm/c;

- average rates 10^{-5} mm/c $< \delta a/\delta N < 10^{-2}$ mm/c;

- high rates 10^{-2} mm/c $< \delta a/\delta N < 1$ mm/c.

As these parameters are set, the test results now have to be found (which is not always that easy!) or we need to carry out some tests within the domain considered and calculate the coefficients that perfectly match these results. In this case, the least squares method is useful to use.

6.4.2. *Simple and constant cyclic loading*

As the frequency does not greatly influence the growth of fatigue cracks. The external loading is defined by:

– its amplitude $\Delta\lambda$ = constant;

– its mean/average value:

$$R = \frac{\lambda_{min}}{\lambda_{max}} = Const \qquad [6.40]$$

The cracking variables only depend on the crack length, with the stress intensity factor in mode I, for instance, being:

$$\Delta K = K_r(a)\Delta\lambda$$

$$R = \frac{\lambda_m}{\lambda_M} \qquad [6.41]$$

These functions have to be used within the cracking law and the corresponding nonlinear differential equation then has to be resolved. If function $K_r(a)$ is not a remarkable function or is only numerically known, a numerical integration has to be performed:

– for Paris' law:

$$\frac{\delta a}{\delta N} = C_1 [\Delta K]^{\eta 1} \qquad\qquad [6.42]$$

with the following initial condition $a = a_0 \to N = 0$:

$$N = \frac{1}{C_1 \Delta \lambda^{\eta_1}} \rightleftharpoons \int_{a0}^{a} \frac{\delta a}{K_r(a)^{\eta_1}} \qquad\qquad [6.43]$$

In the particular case where $K_r(a)$:

$$K_r(a) = \Lambda \sqrt{a}$$

$$\Lambda = Const$$

$$\eta_1 \neq 2 \qquad\qquad [6.44]$$

$$a = \left[a_0^{1 - \eta_1/2} + \left[1 - \frac{\eta_1}{2} \right] C_1 \Delta \lambda^{\eta_1} \Lambda^{\eta_1} N \right]^{\frac{1}{1 - \eta_1/2}}$$

– the other laws can be treated using a step-by-step integration:

$$N_i$$
$$\downarrow$$

$$a_i$$

$$N_{i+1}$$
$$-$$

$$[6.45]$$

$$a_{i+1} = a_i + \delta a_i$$

$$\delta a_i = f(K_r(a_i)\ \Delta\lambda, R)$$

Some examples of the prediction of crack growth compared to the experiments (dotted line) are given in Figure 6.13. They correspond to the three cases shown in Figure 6.12. For these examples, which are still quite simple, the theory-experiment gap, given as the number of cycles for a given crack length, does not go over 10–20%, which is of the magnitude order of the expected prediction.

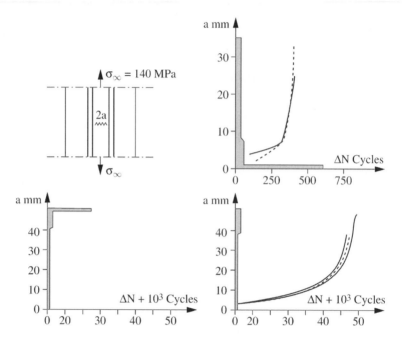

Figure 6.13. *Prediction of the crack growth on stiff sheets made from a light 2618A-T6 alloy*

6.4.3. *Loadings with a complex history*

Still regarding the hypothesis of a simple loading, the external forces can vary at every cycle through the scalar variable $\lambda : \lambda(N)$. Numerical integration then has to be used, step-by-step, as previously done. Wether it is the Paris-Wheeler model or the model with the two variables, K_M and K_S, or more complex models, the calculation time is not that different, as long as these models are represented by differential equations.

For the Paris-Wheeler model, we have:

$$N_i$$
$$\downarrow$$
$$a_i$$

$$N_{i+1}$$
$$\downarrow$$
$$a_{i+1} = a_i + \delta a_i$$

[6.46]

$$\delta a_i = C_5 \left[K_r(a)\lambda_{Mi} \frac{1-R_i}{1-mR_i} \right]^{\eta_5} \left[\frac{\rho_{(K_{Mi})}}{\rho_{z(Ni)}} \right]^{\beta_1}$$

with:

$$\rho_{\left(K_{Mi}\right)} = \frac{K_r^2(a_i)\lambda_{Mi}^2}{2\pi\sigma_y^2}$$

$$\rho_{(Ni)} = \frac{K_r^2(a_S)\lambda_{MS}^2}{2\pi\sigma_y^2} - [a_i - a_S]$$

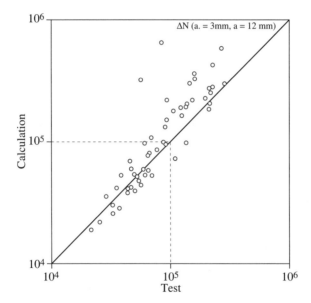

Figure 6.14. *Calculation-test comparison for Falstaff's loading history*

Figure 6.14 represents a set of predictions, compared to the experimental results, for crack growths on loaded sheets in mode I by the loading histories. The loading histories are simulated by using the Falstaff procedure [VAN 75]. The forces are those undergone in the attachment area of an aircraft's wing made of a 2124-T351 alloy during different ground-air-ground cycles (see Figure 6.15). Calculation is performed with the model bearing the two variables, K_M and K_S, integrated cycle by cycle. Coefficient β_2 is adjusted on a specific series of ground-air-ground cycles (see Figure 6.16) [PEL 77a, PEL 77b, PEL 77c].

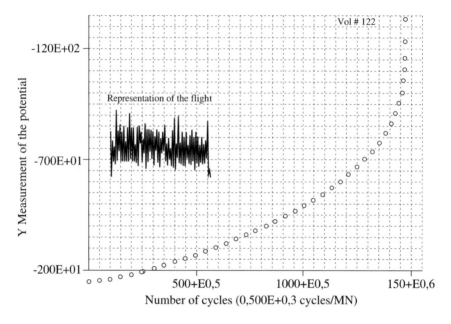

Figure 6.15. *Example of the Falstaff ground-air-ground cycle*

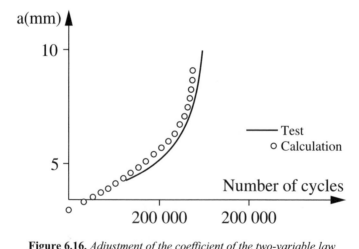

Figure 6.16. *Adjustment of the coefficient of the two-variable law on a Falstaff loading*

6.4.4. *Random loading*

If only the statistics of the loading are known – for instance, by using the probability density of the amplitude of the external forces – the statistic magnitudes related to $a(N)$ cannot be directly calculated. The response of the strongly nonlinear systems (which is the case of the cracking models) to some random inputs is still an unresolved issue [LEM 71]. Two possibilities remain:

– the use of the *RMS* laws [ROL 77] or simulation tests [SCH 75];

– the use of a method based on the Monte Carlo method [PEL 75]. This consists of performing the following numerical steps:

- knowing the probability density p related, for example, to $\Delta\lambda$. Some representative samples of $p(\Delta\lambda)$ are created by haphazardly choosing a succession of values of $\Delta\lambda$ taken from a table $p(\Delta\lambda)$. Let there be m samples (of the order of about 50 to 100):

$$\Delta\lambda_{1i}(N_i), \Delta\lambda_{2i}(N_i),..., \Delta\lambda_{mi}(N_i) \tag{6.47}$$

- the cracking equation is resolved for these m cases (as done previously) as $\Delta\lambda_i(N_i)$ is known, which then comes to:

$$a_{1i}(N_i), a_{2i}(N_i),..., a_{mi}(N_i) \tag{6.48}$$

- averages and standard-values can then be calculated, for instance on the crack lengths, in order to obtain the given statistical magnitudes:

$$\overline{a_i}\left(N_i\right) = \frac{1}{m}\sum_{j=i}^{m} a_{ji}\left(N_i\right) \tag{6.49}$$

The main drawback of this method is the large amount of computing time necessary. A prediction on 100,000 cycles with 50 samples requires δa to be calculated 5 million times!

6.4.5. *Non-simple loading*

If the direction of loading varies during the predicted lifetime of the structure, a bifurcation criterion has to be applied each time that vector $\overrightarrow{P(M)}$ varies.

The prediction class within the problems of the growth prediction of fatigue cracks within structures is about 10–50% of the value of the number of cycles corresponding to a given crack length. This can be minimized by making sure that the best prediction determines failure issues using a conception of the structures based on the fracture mechanics before the project starts [LIE 69a, LIE 69b].

6.5. New directions

There are a number of new directions that crack growth prediction is moving in. Some parameters representing physical mechanisms in the cracking law have been introduced. Cracking laws under complex loading histories have been determined. We now know how to calculate the variable ruling the behavior of the surface cracks in three-dimensions. Crack growth laws under 3D and of metals at high temperature under visco-plasticity conditions have been elucidated.

We have observed the probabilistic property of the loadings and initial defects within the structures.

Finally, we have produced the calculation methods of simplified structures.

6.6. Bibliography

[BUI 78] H.D. BUI, *Mécanisme de la Rupture Fragile*, Masson, Paris, 1978.

[BOI 77] J.M. BOISSENOT, M. DUBOIS, "Study of directional criteria of crack instability", *Journal de Mécanique Appliquée*, vol. 1, no. 2, pp. 133-157, 1977.

[BOI 77] J.M. BOISSENOT, "Analyse tridimensionnelle de la propagation en fatigue d'un défaut de forme semi-elliptique", *Symposium IAEA SM*, pp. 133-157, Vienna, 1977.

[BRO 74a] D. BROEK, *Elementary Engineering Fracture Mechanics*, Nijhoff, Dordrecht, 1974.

[BRO 74b] D. BROEK, "The prediction of crack propagation", *Fracture Mechanics of Aircraft Structures*, AGARD AG-176, pp. 167-180, 1974.

[BUE 70] H.F. BUECKNER, "A novel principle for the computation of stress intensity factors", *Z. Angew Math. Mech.*, vol. 50, pp. 529-546, 1970.

[CHA 74] J.L. CHABOCHE, A. MONTHULET, "Calcul du facteur d'intensité des contraintes pour la prévision de la progression des fissures", *Rech. Aérospat.*, pp. 221-237, 1974.

[CRA 63] S.H. CRANDALL, *Random Vibration*, vol. 2, MIT Press, London, 1963.

[ELB 71] W. ELBER, "The significance of fatigue crack closure", *ASTM STP*, vol. 486, p. 230-242, 1971.

[ERD 63] F. ERDOGAN, G. SIH, "On the crack extension in plate under plane loading and transverse shear", *Journ. Basic. Eng.*, vol. 85, pp. 519-527, 1963.

[ESC 78] Y. D'ESCATHA, "Remarque sur deux critères de rupture fragile pour les problèmes tridimensionnels en mode I", *Journal de Mécanique Appliquée*, vol. pp. 541-552, 1978.

[GIL 78] P. GILLES, Etude des chemins de fissuration dans les structures planes soumises à des chargements complexes, PhD Thesis, Paris VI, 1978.

[GRI 20] A.A. GRIFFITH, "The phenomena of rupture and flow in solids", *Phil. Trans. Roy. Soc.*, Series A, vol. 221, pp. 163-197, 1920.

[HEL 75] J.K. HELLEN, "On virtual crack extensions. On the method of virtual crack extension", *Int. J. Num. Meth. Eng.*, vol. 9, no. 1, pp. 187-207, 1975.

[KAS 75] M.K. KASSIR, G.C. SIH, *Mechanics of Fracture (2): Three-Dimensional Crack Problems*, Noordhoff, Leyden, 1975.

[LAB 78a] R. LABOURDETTE, J. PELLAS, "New approach to the three-dimensional crack growth problem", *Int. J. Fracture*, vol. 14, pp. R121-R124, 1978.

[LAB 78b] R. LABOURDETTE, J. PELLAS, "Energy release rate computation by an inverse flexibility method", *Eng. Fract. Mech. J.*, vol. 10, pp. 879-885, 1978.

[LAC 76] J.C. LACHAT, J. WATSON, "Effective numerical treatment of boundary integral equations. A formulation for 3-D elastostatics", *Int. J. Num. Meth. Eng.*, vol. 10, pp. 991-1005, 1976.

[LEM 71] J. LEMAITRE, La réponse des systèmes non linéaires aux excitations aléatoires, N.T. ONERA, pp. 186, 1971.

[LEM 76] J. LEMAITRE, "Extension de la notion de taux d'énergie de fissuration aux problèmes tridimensionnels et non linéaires", *C.R. Acad. Sc. Paris*, vol. 282, pp. 157-160, 1976.

[LEM 77] J. LEMAITRE, A. TURBAT, R. LOUBET, "Fracture mechanics analysis of pressurized cracked shallow shells", *Eng. Fract. Mech. J.*, vol. 9, pp. 443-460, 1977.

[LEM 78] J. LEMAITRE, "Aspect phénoménologique de la rupture par fissuration", *Mechanics Seminar*, University of Paris VI, 1978.

[LIE 68a] H. LIEBOWITZ, *Fracture II. Mathematical Fundamentals*, Academic Press, London, 1968.

[LIE 68b] H. LIEBOWITZ, *Fracture I. Microscopic and Macroscopic Fundamentals*, Academic Press, London, 1968.

[LIE 69a] H. LIEBOWITZ, *Fracture IV. Engineering Fundamentals and Environmental Effects*, Academic Press, London, 1969.

[LIE 69b] H. LIEBOWITZ, *Fracture V. Fracture Design of Structure*, Academic Press, London, 1969.

[LIE 69c] H. LIEBOWITZ, *Fracture VI. Fracture of Metals*, Academic Press, London, 1969.

[LIE 71] H. LIEBOWITZ, *Fracture III. Engineering Fundamentals and Environmental Effects*, Academic Press, London, 1971.

[LIE 72] H. LIEBOWITZ, *Fracture VII. Fracture of Non Metals and Composites*, Academic Press, London, 1972.

[LUX 77] A.R. LUXMOORE, D.R.J. OWEN, "Numerical methods in fracture mechanics", *1st Int. Congr.*, Pineridge Press, Swansea, 1977.

[NGU 78] Q. SON NGUYEN, "Sur l'utilisation des critères de l'énergie en rupture et en fatigue", *C.R. Acad. Soc.*, vol. 286, Series A, pp. 483, 1978.

[PEL 75] J. PELLAS, J. LEMAITRE, Prévision numérique de la progression des fissures de fatigue sous charge aléatoire, R.T. ONERA, 6/5121 RY, 1975.

[PEL 77a] J. PELLAS, G. BAUDIN, M. ROBERT, "Fatigue crack growth model prediction with two differential constitutive equations", *I.C.F. 4 Proceedings*, pp. 1353-1360, 1977.

[PEL 77b] J. PELLAS, G. BAUDIN, M. ROBERT, "Mesure et calcul du seuil de fissuration après surcharges", *Recherche Aérospatiale*, vol. 3, pp. 191-204, 1977.

[PEL 77c] J. PELLAS, "Lois de progression de fissure de fatigue. La rupture. Sa prévision par calcul", *CETIM.*, Senlis, 1977.

[ROL 77] S.T. ROLFE, J.M. BARSOM, *Fracture and Fatigue Control in Structures*, Prentice-Hall, Englewood Cliffs, NJ, 1977.

[SCH 73] J. SCHIJVE, "Effect of load sequences on crack propagation under random and program loading", *Eng. Fract. Mech. J.*, vol. 5, pp. 269-280, 1973.

[SCH 75] J. SCHIJVE, *Review of Aeronautical Fatigue Investigation in the Netherlands During the Period 1973-75*, NLR MP, 1975.

[SIH 73] G.C. SIH, *Mechanics of Fracture (1): Methods of Analysis and Solution of Crack Problems*, Noordhoff, Leyden, 1973.

[SIH 74] G. SIH, "Strain energy density factor applied to mixed mode crack problems", *Int. J. Fracture*, vol. 10, pp. 305-321, 1974.

[SIH 77a] G.C. SIH, *Mechanics of Fracture (3): Plates and Shells with Cracks*, Noordhoff, Leyden, 1977.

[SIH 77b] G.C. SIH, *Mechanics of Fracture (4): Elastodynamic Crack Problems*, Noordhoff, Leyden, 1977.

[STR 74] M.C. STRIFORS, "A generalized force measure of conditions at crack tips", *Int. J. Solid Structures*, vol. 10, pp. 1389-1404, 1974.

[VAN 75] G.M. VAN DIJK, J.B. DE JONGE, Standard for Fatigue Evaluation Falstaff, NLR MP 75017 U, 1975.

[WEI 76] R.P. WEI, R.I. STEPHENS, "Fatigue crack growth under spectrum loading", *ASTM STP*, 595, 1976.

[WHE 72] O.E. WHEELER, "Spectrum loading and crack growth", *J. Basic. Eng.*, No. 94, pp. 181-186, 1972.

[WIL 71] J. WILLENBORG, R.M. ENGLE, H.A. WOOD, A Crack Growth Retardation Model using an Effective Stress Concept, AFFDL TM 71.1, 1971.

[ZIE 73] O.C. ZIENKIEWICZ, *La Méthode des Éléments Finis*, Ediscience, Paris, 1973.

List of Authors

Stephen D. ANTOLOVICH
Georgia Institute of Technology
Atlanta
USA

Claude BATHIAS
Paris X
Ville d'Avray
France

Gregory GLINKA
University of Waterloo
Canada

Jean LEMAITRE
LMT
ENS Cachan
France

André PINEAU
Centres des matériaux
ENSMP
Evry
France

Bruno SUDRET
LaMSID
EDF R&D
Clamart
France

Index